New Roads and Street Works Act 1991

Code of Practice for the Co-ordination of Street Works and Works for Road Purposes and Related Matters

Second Edition

Approved by the Secretary of State for the Environment, Transport and the Regions under sections 56, 59 and 60 of the Act

April 2001

Department of the Environment, Transport and the Regions

London: The Stationery Office

Department of the Environment, Transport and the Regions
Eland House
Bressenden Place
London SW1E 5DU
Telephone 020 7944 3000
Internet service http://www.detr.gov.uk/

© Crown Copyright 2001

Copyright in the typographical arrangement and design vests in the Crown.

Extracts of this publication may be made for non-commercial in-house use, subject to the source being acknowledged.

Applications for reproduction should be made in writing to The Copyright Unit, Her Majesty's Stationery Office, St Clements House, 1-16 Colegate, Norwich NR3 1BQ.

ISBN 0 11 552310 3

Printed in Great Britain on material containing 75% post-consumer waste and 25% ECF pulp.

TJ 003692   C35   03/01   5673

Contents

Foreword   7

Executive Overview   9

    A   Application of this Code
    B   Executive Overview to the First Edition
    C   Key Changes

INTRODUCTION   15

CHAPTER 1   17

Scope

CHAPTER 2   19

Co-ordination — The Statutory Background

CHAPTER 3   21

Streets subject to Special Controls

    3.1   General   21
    3.2   Protected Streets   22
    3.3   Streets with special engineering difficulties   24
    3.4   Traffic-sensitive streets   28
    3.5   Restrictions following substantial road works   32
    3.6   Directions under section 56   34
    3.7   Section 85 Notices   36
    3.8   Other features of the street   37

CHAPTER 4   39

The Register

    4.1   Background   39
    4.2   Contents   39
    4.3   Additional Street Data   40
    4.4   Inspection of the Register   40

CHAPTER 5 — 43

Work Categories and Notice Requirements

- 5.1 Introduction — 43
- 5.2 Emergency works — 46
- 5.3 Urgent works — 48
- 5.4 Minor works — 50
- 5.5 Standard works — 52
- 5.6 Major projects — 54
- 5.7 Works in streets subject to special controls — 55
- 5.8 Notice validity — 59
- 5.9 Restrictions following substantial road works — 60
- 5.10 Street authority's works for road purposes — 61
- 5.11 Noticing Procedure — Trench Sharing — 61
- 5.12 Other Authorities — 62

CHAPTER 6 — 63

Co-ordination in Action

- 6.1 Information — the key to co-ordination — 63
- 6.2 Co-ordination machinery — 64
  - Regional HAUC
  - Local Co-ordination
  - Terms of Reference
- 6.3 Licensees — 65
- 6.4 Liaison with other bodies — 66

CHAPTER 7 — 67

Related Matters

- 7.1 Location of works and apparatus – exchange of information — 67
- 7.2 Prospectively maintainable highways — 67
- 7.3 Road closures and traffic restrictions — 68
- 7.4 Street works licences — 71
- 7.5 Maintenance of undertakers' apparatus — 72
- 7.6 Obstructions and delays — 74
- 7.7 Reinstatements — 74

CHAPTER 8 — 77

Charges for Unreasonably Prolonged Occupation of the Highway

- 8.1 Basic Concepts — 77
- 8.2 Duration of Works — 78
- 8.3 New Notices — 79
- 8.4 Charging where Works are Unreasonably Prolonged — 83
- 8.5 Notices for Works on a Junction — 84

CHAPTER 9   **87**

Conciliation and Arbitration

    9.1   Introduction   87
    9.2   Conciliation   87
    9.3   Arbitration   88

APPENDIX A   **89**

Definitions

APPENDIX B   **97**

Timeline Diagrams & Flow Charts

APPENDIX C   **115**

Sample Form for Section 58 Notice

APPENDIX D   **117**

Paper Transfer of Notices

APPENDIX E   **123**

Electronic Transfer of Notices

APPENDIX F   **155**

Street Works at or near Level Crossings

APPENDIX G   **169**

Street Works near Highway Structures

APPENDIX H   **189**

HAUC Terms of Reference

FOREWORD

The New Roads and Street Works Act 1991(NRSWA), also referred to in the body of the Code as "the Act" or "NRSWA", supported by relevant Regulations and Codes of Practice, provides a legislative framework for street works activities by undertakers (including public utilities), and also for works for road purposes, to the extent that these must be co-ordinated with street works. The aim is to balance the respective statutory rights of highway authorities and undertakers to carry out works in the street or highway, against the rights of road users to expect the minimum disruption from these works. In particular, the Secretary of State for the Environment, Transport and the Regions is empowered to issue or approve codes of practice giving practical guidance to street authorities and undertakers under the following sections of NRSWA to issue or approve codes of practice giving practical guidance to street and road works authorities:

(a) sections 56(4) — in relation to the power of authorities under those sections to give directions as to the timing of street and road works;

(b) sections 59(3) — in relation to the duty of authorities under those sections to co-ordinate street and road works;

(c) sections 60(2) — in relation to the duty of undertakers under those sections to co-operate with authorities and other undertakers.

This Code of Practice has been approved by the Secretary of State under his powers in the 1991 Act, and provides practical guidance on a wide range of responsibilities of authorities and undertakers in relation to street or highway works. In addition to those responsibilities for which the Act specifically provides for the use of a Code of Practice, the Code also incorporates the requirements set out in a number of sets of regulations relating to notification and registration of works, and the levying of charges upon undertakers when their occupation of the highway is considered to be unreasonably prolonged. The Secretary of State is also empowered under the 1991 Act to make Regulations relating to registers, notices, directions and designations. The following sets of regulations have been made under these powers and their requirements incorporated in this Code.

- The Street Works (Registers, Notices, Directions and Designations) Regulations 1992, S.I. 1992 No. 2985;

- The Street Works (Registers, Notices, Directions and Designations) (Amendment) Regulations 1995, S.I. 1995 No. 990;

- The Street Works (Registers, Notices, Directions and Designations) (Amendment No. 2) Regulations 1995, S.I. 1995 No. 1154 (*revoked*);

- The Street Works (Registers, Notices, Directions and Designations) (Amendment No. 3) Regulations 1995, S.I. 1995 No. 2128;

- The Street Works (Registration Fees) Regulations 1999, S.I. 1999 No. 1048;

- The Street Works (Registers, Notices, Directions and Designations) (Amendment) Regulations 1999, S.I. 1999 No. 1049;

- The Street Works (Charges for Unreasonably Prolonged Occupation of the Highway) (England) Regulations 2001.

This Code was prepared by a working party of the Highway Authorities and Utilities Committee (including representatives of the National Joint Utilities Group, the Local Government Association, and the Department of the Environment, Transport and the Regions) and was the subject of extensive consultation with relevant organisations. This Code applies to England only and replaces the original code that was issued in November 1992. It will come into operation on 1 April 2001. The new code does not apply to Scotland and Wales, which will continue, for the time being, to be subject to the 1992 code.

Department of the Environment, Transport and the Regions

February 2001

EXECUTIVE OVERVIEW

A Application of this Code

1 The first edition of this Code of Practice came into force in England, Wales and Scotland on 1 April 1993. Much has been learned about co-ordination since then, and this edition of the Code has been significantly revised in the light of experience. An outline of the main changes can be found at Section C below. However, the key principles underlying the Code have not changed at all and therefore the original Executive Overview is as applicable now as it was then. It is reproduced in full, with minor amendments to reflect changes with the passage of time, in Section B.

One major change since the first edition is devolution. Whereas the first edition of this Code applied to England, Wales and Scotland, at the time of going to print this second edition applies to England only. A further complication is that, between editions, a stand-alone document promulgating a new Appendix (Appendix E) came into force in England and Wales only, on 1 April 1999. Scotland was not included because Appendix E deals with the electronic transfer of notifications and Scotland already had its own system in place when the Appendix was issued. Appendix E has also been revised in the light of experience and the decision to implement the terms of section 74 of the Act. The opportunity has been taken to incorporate the revised Appendix E into the same binding as the rest of the Code.

The situation in the three jurisdictions is therefore as follows:

Scotland   The First Edition (1/4/1993) of this Code applies;
Wales   The First Edition (1/4/1993) plus the stand-alone Appendix E (1/4/1999) apply;
England   This Second Edition (1/4/2001) of the Code applies.

B Executive Overview to the First Edition

1 One of the most important elements of the street works legislation is the duty on street authorities to co-ordinate all works in the highway. As important is the parallel duty on undertakers to co-operate in this process. It is essential that both street authorities and undertakers take these responsibilities seriously. This Code of Practice is intended to help them do so.

2 The New Roads and Street Works Act 1991 ("the Act") spells out the objectives of the co-ordination function. They are:

— to ensure safety;

- to minimise inconvenience to people using a street, including a specific reference to people with a disability;

- to protect the structure of the street and apparatus in it.

It is essential that everyone responsible for planning and carrying out works in the highway take these objectives into account.

3   The legislative and regulatory provisions explained in the Code are perhaps best viewed as three pillars of the co-ordination framework:

- the notice system - the notices themselves provide vital information to aid the co-ordination process, while the notice periods provide time within which appropriate steps can be taken. The best method of exchanging notices is electronically, as provided in Appendix E to this Code;

- streets subject to special controls — these designation procedures provide a mechanism by which attention can be focussed on particularly sensitive streets. Particularly important in this context are traffic-sensitive streets;

- the co-ordination tools — the legislation provides a range of tools to facilitate the co-ordination process. These include: the power to direct the timing of street works; the power to restrict street works following substantial road works; and the requirement on undertakers to avoid unnecessary delay and obstruction.

4   There are three key principles to which undertakers and street authorities must adhere if this aspect of the legislation is to be effective. They are:

- the need to balance the potentially conflicting interests of road users and undertakers' customers. All concerned must remember that the motorists and pedestrians who suffer as a result of works of all kinds in the street are the same people who benefit from and depend on a reliable service from the undertakers;

- the importance of close co-operation and liaison between street authorities and undertakers. Good communication is central to the co-ordination process;

- an acknowledgement on all sides of the fact that works programmes and practices may have to be adjusted in order to ensure that the statutory objectives of the co-ordination provisions are achieved.

5   The Code of Practice explains how these principles can be implemented in practice. Particularly significant elements of the guidance are:

- the fact that the notice periods specified in the legislation and Regulations are minimum periods: wherever possible longer periods of notice should be given and information contained in notices should be updated whenever appropriate;

- the central role played by local liaison meetings between street authorities, undertakers and other interested parties;

— the importance of the designation of streets subject to special controls. Street authorities should ensure that designation is confined to those cases where it is strictly necessary. Undertakers must respect the objectives of designation when planning and carrying out works in such streets.

6 In carrying out their responsibilities under the legislation, street authorities and undertakers should endeavour to ensure that their works are planned in such a way as to minimise inconvenience to all road users including organisations representing disabled people. This has implications for: the timing of works, the way in which they are carried out, and programming of undertakers' and street authorities' medium- and long-term works. Inevitably, in all but the quietest situations, works in the street must interfere with traffic to some extent; the aim should, however, be to ensure that only in exceptional circumstances will:

— serious traffic disruption be caused by street works;

— recently resurfaced or reconstructed streets be affected by street works;

— one undertaker carry out planned street works in a street within a short time of earlier works being carried out.

The street authority is generally the first point of contact for public complaints. In order to ensure the public is aware who is responsible for the works, it is therefore important that the promoting undertaker gives advance information and warning to affected frontagers not only about any disruption to their services, but also if their access is to be affected for any length of time. Details of alternative access arrangements, or any other form of mitigating action to be taken by the promoting undertaker, should be supplied.

7 In no provision is the crucial balance between highway and utility interests better illustrated than the execution of urgent works on traffic-sensitive streets. The concept of urgent works reflects the crucial importance of some types of street works that fall short of emergencies as defined in the Act. Traffic-sensitive streets are those on which street works are most likely to cause traffic disruption.

8 It is important that, in this area in particular, authorities and undertakers work closely together. If they do not the urgent works provisions will fall into disrepute and the credibility of the legislation and the two key players in its implementation will be undermined.

9 Street authorities must pay particular attention to the need to minimise the impact of works on people with disabilities. Similarly, undertakers must take full account of the interests of people with disabilities when planning and implementing their works.

C Key Changes

1 The opportunity has been taken to improve and amplify the text in many places throughout the Code. There are too many changes to list all of them here, but the major changes are explained below.

2   The introduction of the electronic exchange of notices, when Appendix E came into force in England and Wales on 1 April 1999, was a major step forward. It provided a standard mechanism for undertakers to use to send their notices to highway authorities. As well as being far more efficient than previous paper-based systems, electronic interchange has the potential to greatly improve co-ordination by speeding up the communication between undertakers and highway authorities.

3   This edition of the Code incorporates Appendix E, superseding the stand-alone document (in England only, see A above) and Appendix E has been changed in the following ways:

— there are numerous changes throughout the Appendix aimed at correcting some previous ambiguities and generally improving the document;

— the "return path" has been specified. That is, it is intended that highway authorities should, from 1 April 2001, be able to return comments to undertakers electronically. They should also be able to initiate an exchange in certain circumstances, which are:

- when seeking to find an owner for unattributable works; and
- giving section 58 and 85 notices;

— the necessary information to enable the exchange of notices required by the implementation of section 74 of the Act has been inserted (see paragraph 7 below);

— the data sets and data validation required by the Inspections regime have been removed from the Appendix and will appear in the next edition of the Inspections Code of Practice.

4   Section 74 of the Act (*"Charges for the occupation of the highway where works unreasonably prolonged"*) is being implemented on 1 April 2001. The new Chapter 8, of this Code explains the principles behind section 74 and how they will be operated.

The key concepts are:

— the prescribed period: a period prescribed in regulations; and

— the reasonable period: the period that the undertaker estimates that the works will take (if not challenged by the highway authority), or the period agreed with the highway authority, or (if not agreed) the period determined by arbitration (pending the outcome of which the highway authority's view of the reasonable period is acted upon).

— A highway authority may levy a charge if the duration of a works exceeds both of these periods, which effectively means if it exceeds the longer of the two periods.

— New notices have had to be defined to enable section 74 of the Act to operate effectively. These are: Actual Start Date of Works; Revised Duration Estimate; Challenge to Duration Estimate; Works Clear and Works Closed. Provision has been made in Appendix E to enable all these notices to be exchanged electronically.

5   The introduction of electronic noticing brought with it the use of definitive street gazetteers against which to locate street works. The gazetteers also hold information on streets subject to special controls. The opportunity has been taken in this revision to

enable the inclusion on the gazetteers of other, useful information that the highway authority already holds or may easily obtain.

6   This additional street information includes:

— the fact that there are restrictions in place (under section 58 of the Act) following substantial road works;

— the fact that a notice under section 85(2) of the Act has been given which may affect future diversionary works;

— other features of the street, including:

- Environmentally Sensitive Areas;
- Special Surfaces;
- Pipelines;
- Priority Lanes; and
- Special Construction Needs.

7   All of this information will be an aid to the better planning of works and to their better co-ordination in execution. Highway authorities are therefore encouraged to use these new facilities and to include such information as they have available as they update their gazetteers.

8   Two other features of the street, which may be recorded in the gazetteer in the future, are worthy of special mention:

— Structures (not designated as being of Special Engineering Difficulty). These are structures associated with the highway which require extra care being taken when working nearby. The new Appendix G gives guidelines to be followed when working near these structures; and

— Level Crossing Precautionary Areas. When works are proposed or executed in these areas around a level crossing, the extra precautions detailed in the new Appendix F must be applied.

9   Appendix B contains some revised flow charts plus some new time-line diagrams intended to illustrate the operation of section 74 of the Act.

10 Appendix D contains a changed format for paper notices where these are still used. As the vast majority of practitioners are now using electronic notices, the opportunity has been taken to revise the paper notice to bring it into line with electronic notices. This has many advantages, including obviating the need to devise new paper notices for section 74 and ensuring that highway authorities receive all the information they need (such as Unique Street Reference Numbers (USRNs)) on paper notices as well as in electronic ones.

INTRODUCTION

1. The New Roads and Street Works Act 1991 was enacted following the comprehensive review of the Public Utilities Street Works Act 1950 by a Committee chaired by Professor Michael Horne. The Committee's report was presented to the Government in 1985.

2. This Report highlighted the unnecessary traffic disruption and significant additional costs to the public (apart from those incurred by undertakers, highway authorities and transport operators) caused by the uncoordinated activities of those concerned in carrying out works of all types on the highway.

3. It is incumbent on everyone working within the highway, whether street authority, undertaker, private developer or builder, to take into account the needs of all road users including those with disabilities, whether they be pedestrian, cyclist or driver, at all stages in the planning and execution of works in the street. Consequently, a positive attempt has to be made to reconcile all the differing interests. Accordingly, one of the features of the Act was the creation of specific obligations for both street authorities and undertakers with respect to co-ordination and co-operation (under sections 59 and 60).

4. The Act also established a notices and registration regime, and gave street authorities powers to issue directions, in specified circumstances, as to the time when street works may or may not be carried out. For the purposes of the statutory notice requirements, provision was also made for streets to be designated as traffic-sensitive, in accordance with prescribed criteria. Existing powers, which restricted the execution of street works following resurfacing or reconstruction of the street, were re-enacted in an amended form. The law relating to special roads was extended to provide a class of "protected street", and a further category of street was created to safeguard sensitive engineering features.

5. This Code of Practice on the implementation of the Act and the Regulations with respect to co-ordination, notices and registration and related matters provides guidance to all concerned as to the manner and circumstances in which these various provisions should be operated.

6. This second edition of the Code is the first revision, and includes information and advice on the operation of the newly implemented section 74 of the Act, as well as other new material. It applies only to works carried out in England.

7. The requirements of this Code apply to all who carry out works on the highway, as appropriate and indicated in the text.

8. The Appendices, which follow the main body of the text, are as follows:

    Appendix A provides definitions of terms used in the Code and not defined elsewhere;
    Appendix B provides Timeline Diagrams and Flow Charts;
    Appendix C provides a Sample Form for the notice of Substantial Road Works;
    Appendix D describes the Paper Transfer of Notices;
    Appendix E describes the Electronic Transfer of Notices;
    Appendix F describes new procedures for dealing with Railtrack level crossings;
    Appendix G sets out procedures for working near highway structures;
    Appendix H sets down the Terms of Reference for regional HAUCs and Local Co-ordination meetings.

CHAPTER 1
Scope

1.1 This Code of Practice has been approved by the Secretary of State under sections 56, 59 and 60 of the Act. It gives practical guidance to street authorities on the exercise of their powers to give directions as to the timing of street works and to street authorities and undertakers as to the discharge of their duties with respect to co-ordination and co-operation.

1.2 In addition to the sections referred to above, the Code also includes advice on non-statutory agreements made by the Highway Authorities and Utilities Committee (HAUC), all of which are central to the effective co-ordination of works in the street. These additional agreements relate to the following provisions, which are also referred to in the Code:

Section 50	—	(Street Works Licences)
Section 53	—	(The Street Works Register)
Sections 54 & 55	—	(Regarding advance notice and notice of commencement of works)
Section 58	—	(Restrictions following substantial road works)
Sections 61 to 64	—	(Streets subject to special controls)
Section 66	—	(Avoidance of unnecessary delay and obstruction)
Section 74	—	(Charge for occupation of the highway where works are unreasonably prolonged)

1.3 The Code summarises the statutory requirements as to notice periods, notice formats and methods of service. It also lays down the procedures to be adopted for the designation of Protected Streets, Streets with Special Engineering Difficulties and Traffic-Sensitive Streets, and when works are proposed in them, and the circumstances in which directions may be given as to the timing of street works.

1.4 The Code provides agreed procedures for conciliation and arbitration where disputes or differences arise.

1.5 In a brief but most important chapter (Chapter 6), the Code pulls together the various strands in a note devoted to the co-ordination function itself, giving practical advice for all concerned on how the objectives may best be achieved.

CHAPTER 2
Co-ordination — the statutory background

2.1 This chapter sets out the statutory basis for the co-ordination of works of all kinds in the street.

2.2 Section 59(1) of the Act requires street authorities to "use their best endeavours to co-ordinate the execution of works of all kinds (including works for road purposes) in streets for which they are responsible —

(a) in the interests of safety,

(b) to minimise the inconvenience to persons using the street (having regard in particular to the needs of people with a disability), and

(c) to protect the structure of the street and the integrity of apparatus in it."

That duty extends to co-ordination in appropriate cases with other street authorities where works in a street for which one authority are responsible affect streets for which other authorities are responsible. This will, by definition, include street managers, where appropriate.

2.3 Section 60 places a duty on undertakers to use their best endeavours, in regard to the execution of street works, to co-operate with the street authority and one another, with the same threefold objectives mentioned in the previous paragraph.

2.4 These sections provide for this Code of Practice and impose a statutory obligation on street authorities and undertakers to have regard to it in the steps they take to co-ordinate their works.

2.5 In addition to these specific co-ordination provisions, the following sections are all essential components in the machinery of co-ordination. They are noted in greater detail elsewhere in the Code.

Section 53	—	the street works register (Chapter 4)
Sections 54, 55 and 57	—	notices (Chapter 5 Sections 5.2 to 5.6 inclusive)
Section 56	—	directions as to timing (Chapter 3 Section 3.6)
Section 58	—	restrictions on street works (Chapter 3 Section 3.5 and Chapter 5 Section 5.9)
Sections 61 to 64	—	streets subject to special controls (Chapter 3 Sections 3.2, 3.3 and 3.4, and Chapter 5 Section 5.7).

Section 74 — unreasonably prolonged occupation of the highway (Chapter 8).

Section 85 — notice of major highway, bridge or transport works (Chapter 3 Section 3.7)

2.6 The remainder of this Code of Practice explains in more detail the way in which it is intended that these provisions should operate. Chapter 6 will then explain how they can be used as part of an effective co-ordination regime.

CHAPTER 3

Streets subject to special controls

3.1 General

3.1.1   The notice and co-ordination regime is designed to balance the need to restrict to a minimum the bureaucracy involved in managing street works with the importance of minimising delay and inconvenience to road users and protecting the integrity of the street and apparatus within it. The provisions of the Act, which enable certain categories of special streets to be designated, are central to achieving this balance. The three categories of street subject to special controls are:

— protected streets;

— streets with special engineering difficulties;

— traffic-sensitive streets.

3.1.2   The Regulations define the criteria to be adopted in designating these categories of street and the procedures that have to be followed in doing so. The detailed implications of designation for those proposing to carry out street works are explained in Chapter 5 Section 5.7. This chapter will, in the case of each type of street:

— explain the aim of designation;

— explain the criteria to be adopted in designation;

— provide advice on the approach to be adopted by street authorities and undertakers.

In addition to the above, this section gives guidance on:

— restrictions following substantial road works (section 58)

— directions (section 56)

— other features of the street on which information may be contained in the Additional Street Data sets to the National Street Gazetteer (NSG).

3.2 Protected streets

Background

3.2.1 By virtue of section 61 of the Act, all special roads within the meaning of the Highways Act 1980 (e.g. motorways) are protected streets. In addition, a street authority may designate other protected streets meeting criteria specified in the Regulations.

3.2.2 Protected streets may be designated only if they serve or will serve a specific, strategic major traffic need and there is a reasonable alternative route in which undertakers can place the equipment which would otherwise lawfully have been placed in the protected street, e.g. services to existing or proposed properties in the street, or trunk supply routes passing through the street.

3.2.3 Undertakers will be supplied with information on protected streets via the NSG concessionaire.

3.2.4 Chapter 5 Section 5.7 gives advice regarding notice periods and other procedures required in protected streets.

The implications of designation

3.2.5 Once a street has been designated as protected, the activities of undertakers and street authorities will be severely restricted. No undertaker's apparatus may be placed in the street (except by way of renewal) without the street authority's consent (although, under the Act, lateral crossings should normally be allowed). Additionally, undertakers' works in verges and central reservations not impinging on the carriageway should usually be acceptable. Road maintenance or repairs will, in general, only be carried out at night or, where appropriate, at weekends, or at other times when the impact upon traffic will be minimised.

Existing streets

3.2.6 Given the nature of the powers in relation to protected streets and the possible financial implications for both street authorities and undertakers, designation should be contemplated only when absolutely essential. The decision to designate should be taken only after extensive consultation and, in the case of existing streets, after other means of reducing delay and inconvenience caused by street works have been explored.

3.2.7 Where an existing street is being considered for designation, the street authority must justify the need. They must therefore:

(a) take account of the needs of utilities both to supply and maintain services to frontagers and to use such streets for existing trunk supplies;

and

(b) where they require the removal or alteration of apparatus in the street, reimburse all reasonable expenses incurred by the undertaker in so doing (subject to appropriate allowances for betterment, deferment of renewal and value of recovered apparatus). The cost sharing arrangements for diversionary works will not apply.

New streets

3.2.8 Where a newly-constructed street is being considered for designation, the street authority will initially consult with all undertakers who might have an interest, and will, where requested and reasonably practicable, make provision, at the undertaker's expense, for the necessary service areas or service strips alongside carriageways, and for duct or service crossings.

Procedure for making and withdrawing designations

3.2.9 The detailed procedure is laid down in the Regulations. Prior to designating a protected street the street authority must publish a notice of their proposal in a local newspaper and supply a copy of the notice to:

(a) every undertaker whom the authority knows have statutory powers to execute street works in the street,

(b) any local authority (other than the street authority itself) in whose area the street is located,

(c) owners and occupiers of land or premises adjacent to the street,

(d) Transport for London (where appropriate),

(e) any other person who has formally requested to be notified.

If no objection is received within the period of not less than one month mentioned in the notice, the designation may be made. However, where the proposal has led to objections, the street authority must hold a local inquiry into the objection and must consider the inquiry report before deciding whether to proceed with, modify or abandon the proposal.

3.2.10 Whilst the Regulations empower the street authority to withdraw a designation at any time, a designation may also either be withdrawn by agreement or, failing agreement, decided by arbitration. Where the Secretary of State made the original designation, his approval is required before it can be withdrawn. In all cases of withdrawal, the procedure described in the Regulations must be followed.

3.2.11 Designations are recorded in the Additional Street Data sets of the NSG. For details see Chapter 4, Section 4.3.

3.3 Streets with special engineering difficulties

Background

3.3.1 Section 63 and Schedule 4 of the Act refer. The Regulations set out in detail the criteria for designation.

3.3.2 The term 'special engineering difficulties' relates to streets, or, more commonly, parts of streets, associated with structures, or streets of extraordinary construction where works must be specially carefully planned and executed in order to avoid damage to or failure of the road structure with attendant danger to person or property. Plans and sections of proposed works must therefore be approved by each of the relevant authorities, i.e. street authority, or sewer, transport or bridge authority with an interest in the structure concerned.

3.3.3 These are the only circumstances in which a plan and section drawing showing street works proposals is still required so that the authority concerned with the structure may satisfy themselves that no detriment will result from the street works.

3.3.4 Chapter 5 Section 5.7 gives advice regarding procedures necessary in relation to streets with special engineering difficulties.

Scope of designations

3.3.5 The designation of streets with special engineering difficulties should be used only where strictly necessary bearing in mind the safeguards already provided elsewhere in the Act, e.g. sections 69 (for other apparatus in the street), 88 and 89 (for bridges and sewers), and 93 (level crossings and tramways). This is in the interests of all concerned — street authority, undertaker and where appropriate the owner of the structure involved. Advice about bridges and other highway structures can be found in Appendix G.

3.3.6 Circumstances where designation may be appropriate include:

(a) Bridges. The street may be designated if the authority is concerned about the influence of street works on the strength, stability and waterproofing of the bridge, or access for maintaining it, or for any other purpose. In general, the designation would relate to the whole of the bridge structure, but it will only be necessary to designate the area adjacent to the bridge concerned and not the whole length of the street. Three areas may be designated — the bridge deck, an area in the vicinity of the abutments where excavation no deeper than 1.2 metres is appropriate, and another where further restrictions are needed if excavation is deeper.

(b) Retaining walls where they give support to the highway and bridge abutments where the foundations are sufficiently shallow to lead to concern that excavation could affect adversely the integrity of the structure. Where the foundations are piled, designation is only likely to be necessary if the excavation could alter the degree of support given to the piles by the soil.

In many cases, it will only be necessary to designate the area adjacent to the structure concerned and not the whole width of the street. Two areas may be designated — one appropriate for excavations no deeper than 1.2 metres, the other where further restrictions are needed if excavation is deeper.

(c) Cuttings and embankments. Areas adjacent to such structures should be designated if excavation could lead to slides or slips of the soil, or where excavation could affect special construction features such as earth reinforcement systems or lightweight fills. Either the whole width of street may be designated, or specific areas similar to those given in (b) above.

(d) Isolated structures such as high mast lighting columns and large sign gantry supports. Where excavation could affect their stability, areas immediately around the supports should be designated, again distinguishing between excavations up to 1.2 metres deep and those which are deeper.

(e) Subways and tunnels at shallow depth. Areas immediately above the structure and adjacent areas may be designated.

(f) Tramway tracks in the street. Areas occupied by the tracks and immediately adjacent areas may be designated. (Additional protection to the appropriate authority is also given in section 93 of the Act.)

(g) The area of the street immediately above a culvert may be designated where the structural integrity of the pipe or channel could be adversely affected by works. It would not therefore be expected that, for example, a reinforced concrete pipe or box culvert would justify designation. A masonry or steel culvert could be considered if the depth of cover is shallow.

(h) Undertaker's apparatus. Only in exceptional circumstances, such as electricity pylons adjacent to the carriageway, should designation be required.(In most cases the needs for safety and security of apparatus are adequately covered already by the requirements of section 69 or section 89 of the Act.)

(i) The presence of some types of pipeline installed under the Pipelines Act 1962 and similar structures which cross or traverse the street may justify designation.

(j) Streets which pose extraordinary engineering problems in the event of excavation taking place may be designated — for example, a road of weak construction founded on very poor soil, such as a peat bog, which might have utilised geotextiles. (The requirements of the Code of Practice *Specification for the Reinstatement of Openings in Highways*, approved under section 71 of the Act, will provide adequate protection for the vast majority of streets.)

Procedure for making and withdrawing designations

3.3.7 The procedure for making designations is simpler than that for protected streets, and is detailed in the Regulations. Notice in a newspaper is not required; notice must be given to undertakers and other persons who have asked to be given notice, as well as any street authorities concerned, and, in London, Transport for London. The street authority must give a minimum of one month's notice, during which objection may be made.

3.3.8 A relevant authority or undertaker wishing a street to be designated must inform the street authority of the area and length of street involved and supply to the street authority all relevant data relating to the structure concerned (for the accuracy of which he will be responsible) to enable the designation to be made in appropriate cases. If the street authority declines to make the designation, there is a right of appeal to the Secretary of State.

3.3.9 A street authority may also withdraw a designation once it has been made. A relevant authority or an undertaker may apply to them to do so. Where a person other than the owner of the relevant structure makes the request for the withdrawal of designation, it must be supported by appropriate engineering evidence and the street authority should consult the owner of the structure before such a designation is withdrawn. The consent of the Secretary of State is needed where the designation was made on his direction. In all cases of withdrawal the procedure described in the Regulations must be followed.

3.3.10 The street authority shall must record the details of all designations, and withdrawals, in the register. They will do this by using the Additional Street Data sets (ASD) of the NSG (see Chapter 4, Section 4.3). Undertakers will then be able to access these data sets from the NSG.

3.3.11 The information that the street authority will be required to include in the ASD of the NSG must include particulars of the structures involved and whom they belong to, together with the length and area of street involved. Each structure must be identified by its OS grid reference, road identification number, local name etc. and a unique reference number.

Practical considerations

3.3.12 As is clear from paragraph 3.3.5, designations should not be made as a matter of policy wherever there is a bridge or structure which is likely to be affected by street works. Each case should be considered on its merits in the light of the criteria set out in paragraph 3.3.6 above, and the procedures for making or withdrawing any such designations in paragraphs 3.3.7 to 3.3.11. Initial designations that were made on the Act coming into operation in 1993 may therefore on subsequent consideration and in the light of experience since then turn out to be too wide-ranging. Where this is so, street authorities and owners of structures should re-examine the designations concerned with a view to withdrawal of any that are unnecessary in the light of the other safeguards in the Act.

3.3.13 Similarly, undertakers are encouraged to discuss their proposed works with the relevant authority as early as possible. (See also paragraphs 5.7.10 and 5.7.11, and Appendices F & G)

Cellars

3.3.14 It is not practical for the highway authority to identify all cellars under footways and carriageways and to decide whether they justify the designation of the street as one of special engineering difficulties.

However, under section 180 of the Highways Act 1980, owners of cellars wishing to carry out works are required to notify the highway authority, who in turn will notify interested undertakers before any work is carried out.

3.3.16 Highway authorities and undertakers wishing to carry out work in areas where they know, or might reasonably be expected to know, of the existence of cellars should notify the cellar owners or frontagers when they intend to carry out excavations close to cellars or extensive excavations which will impinge upon cellars.

Summary of information to be supplied

3.3.17 The following is a summary of the information that must be supplied by a street authority proposing to designate a section of street as having special engineering difficulties:

(i) A unique reference number of the designation.

(ii) Name of the street authority making the designation.

(iii) Name of the relevant authority responsible for the structure (where it is not the same as (ii) above).

(iv) Type of structure.

(v) Location (e.g. street name, road number, OS grid reference etc).

(vi) Dimensions of the area in which the designation will apply.

(vii) Any other relevant information. (NB: As more information becomes available, it should be issued and, if necessary, the designation reviewed.)

Policy guidance

3.3.18 As stated in paragraphs 3.3.7 to 3.3.11 above, the responsibility for the designation of sections of streets with special engineering difficulties lies with the street authority, which is also responsible for maintaining a list of such designations and notifying undertakers of them. But it is important to appreciate that:

— the owner of the relevant structure must inform the street authority of the existence of the structure so that it can be considered for designation (paragraph 3.3.8); and

— the key relationship, in terms of ensuring that adequate precautions are taken, is that between an undertaker proposing works and the owner of the structure concerned.

3.3.19 If this mechanism is to work effectively, it is essential that:

— there is close co-operation and consultation between the street authority, undertakers, bridge authorities and other owners of relevant structures in relation to the designation and withdrawal of designation of sections of streets with special engineering difficulties;

— arrangements are agreed between undertakers and the owners of structures on the handling of emergency and urgent works on sections of streets with special engineering difficulties;

— in the cases of planned major works and the provision of new supplies there are early discussions between undertakers and the owners of the structures concerned;

— on receipt of formal notices covering street works on sections of street with special engineering difficulties, the street authority ensures that the necessary actions are in hand.

3.4 Traffic-sensitive streets

Background

3.4.1 Under section 64 of the Act a street authority may designate certain streets (or parts of streets) as "traffic-sensitive". The criteria under which such designations are to be made are specified in the 1992 Regulations.

3.4.2 It is expected that these criteria will cater for the great majority of cases. Street authorities will therefore normally apply them as a matter of course. However, to cater for very special local circumstances, such as individual annual or seasonal events or where the existence of the criteria has not been established but there is a danger of street works causing serious disruption to traffic, provision is made for additional designations of traffic-sensitivity by agreement between street authorities and statutory undertakers in their area, i.e. those operating by virtue of a statutory right. There are two important points to remember. Clearly, as designation affects everyone, all undertakers with an interest in the street must be involved in agreeing the designation. Therefore, this type of designation can only be made by agreement with the majority of local statutory undertakers since it is not one which can be determined by arbitration if agreement cannot be reached.

3.4.3 Designation may apply to the carriageway only, or to a footway or pedestrian area only, to part only of a length of street, and to certain times of the day, days of the week or days of the year, depending on the circumstances.

3.4.4 Once a designation is made it applies to all works taking place in the street. Street authorities carrying out works for road purposes as well as undertakers carrying out street works must avoid carrying out those works in traffic-sensitive situations at sensitive times unless there is no alternative.

3.4.5   Chapter 5 Section 5.7 gives advice regarding notice periods and other procedures required in traffic-sensitive streets.

Vehicular routes

3.4.6   Provided they meet the statutory criteria, traffic-sensitive routes may include:

(a) main inter-urban roads and main radial and other commuter routes (many routes near city centres) which carry heavy peak traffic flows;

(b) heavily trafficked routes into and within holiday areas during the holiday/visitor season;

(c) routes that are only traffic-sensitive on certain predetermined occasions; e.g. race meetings, county shows etc.;

(d) routes in London that Transport for London, or their predecessor organisation, the Traffic Director for London, has so designated.

Pedestrian areas

3.4.7   Subject to meeting the criteria these may include:

(a) Commuter routes — footways very heavily used at peak times by commuters;

(b) Special events pedestrian routes — pedestrian routes to special events, e.g. the local showground, football ground etc., which are only sensitive on certain days of the week;

(c) Pedestrian shopping areas — often areas of special paving, with restricted access for delivery vehicles.

Longer notice periods

3.4.8   The Regulations make special provision for longer notice periods where works are to be carried out in a traffic-sensitive situation (i.e. in a designated street at the times to which the designation of traffic-sensitivity applies). This provides an opportunity for the street authority to make any necessary traffic rearrangements.

3.4.9   Longer notice periods also enable the street authority to co-ordinate their own works, or those of other undertakers, to minimise traffic disruption, as well as to agree with the undertaker on the most appropriate times for the works to be carried out, e.g. avoiding peak traffic periods, special events, traffic in pedestrianised areas at congested times, etc. Where agreement on these matters cannot be reached, the street authority may find it necessary to issue directions as to timing of street works under section 56 of the Act (see paragraphs 3.4.19 and Section 3.6 below) to ensure compliance with designations.

Procedure for making and withdrawing designations

3.4.10 The formal procedural requirements of the Regulations (which are similar to those applicable to streets with special engineering difficulties (see section 3.3.7 above) must be satisfied. The street authority must give a minimum of one month's notice, during which objection may be made.

3.4.11 As mentioned in paragraph 3.4.3, designations may be limited to certain times of the day (rush hour periods) or certain dates (known periodic events) where the criteria will only be satisfied at such times or dates. (See section 64(3) of the Act.)

3.4.12 Where a street authority intend to designate a street, or any part of a street, as traffic-sensitive because one or more of the criteria are met, they shall must consult with all interested parties (undertakers, street managers, police, etc.) with a view to reaching agreement on the proposals. Where there is a failure to agree on whether the criteria are satisfied, the designation will stand pending resolution of the difference by conciliation and arbitration. Conciliation must take place within 2 months (as described in Chapter 9), and arbitration, if required, must take place within 10 weeks of the issue being referred to the local (or regional) HAUC.

3.4.13 Where, because of changed circumstances, a designation is no longer appropriate, it should be withdrawn or modified by the street authority, with the agreement of all concerned. An undertaker may in such cases ask the street authority to act. A failure to agree may be referred by either party to the conciliation and arbitration machinery. In the meantime the designation will stand. The Regulations prescribe the formal procedure to be adopted.

3.4.14 As previously described in paragraph 3.4.2, where designations are only subject to agreement because the criteria cannot be met, conciliation and arbitration cannot apply where that cannot be achieved.

3.4.15 Designations are recorded in the Additional Street Data sets of the NSG. For details see Chapter 4 Section 4.3.

Other situations meriting special consideration

3.4.16 Designation of traffic-sensitive situations is intended to apply to locations where serious traffic disruption may ensue from the presence of street works or works for road purposes. There are, however, a number of situations for which designation as traffic-sensitive is not appropriate but where special care needs to be exercised when works are to be carried out.

3.4.17 Situations which might merit special attention (but which would not otherwise satisfy the requirements for traffic-sensitivity) include:

(a) access to busy bus or railway stations;

(b) works at major bus stands on the highway;

(c) carriageway works near to the vehicular access to fire stations, ambulance stations, police stations and hospitals;

(d) footway works impacting on people with visual or other disabilities;

(e) the vicinity of a nurses' home accommodating staff who are likely to be on night shifts;

(f) sites of accident concentration.

3.4.18 It would be appropriate to identify and discuss such situations at the local co-ordination meetings with a view to agreeing means of minimising the occurrence of problems on site. But, where statutory notices are either short or not required, undertakers should discuss their works with local representatives of the organisations referred to in paragraph 3.4.16 above as soon as possible, so as to accommodate their special requirements during execution of the works.

Policy guidance

3.4.19 Undertakers will be supplied with information on traffic sensitive streets via the NSG concessionaire. The credibility of the system will be enhanced if undertakers fully understand the reasons for designation.

3.4.20 The precise steps to be taken when street works are carried out on traffic-sensitive streets will vary from case to case and should be the subject of discussion at the regular co-ordination meetings (see Chapter 6). In most cases arrangements should be based on the outcome of those discussions, but in default of agreement a street authority may use their powers to direct the timing of street works under the provisions of section 56 of the Act (see Section 3.6 below).

3.4.21 By definition traffic-sensitive streets are those on which delays and inconvenience to road users must be kept to a minimum. To a great extent the success of this legislation will be judged in terms of its impact on these streets. This means that:

— in order to ensure that such streets receive the attention they deserve, street authorities should seek to designate only those streets on which the extra controls provided by designation are fully justified;

— undertakers should ensure that they respect the objectives of designation from the outset in planning and implementing work on such streets.

Summary of information to be supplied

3.4.22 The following is a summary of the information which must be supplied by a street authority proposing to designate a street as traffic-sensitive:

(i) name of street authority making the designation;

(ii) location of the street or part of the street to be designated (e.g. the name or number of the street, identification of sections between junctions etc);

(iii) times when the designation will apply (i.e. times of the day, days of the week, days of the year or a particular date upon which the designation applies);

(iv) justification for the designation (i.e. either by satisfying the traffic flow criteria or as agreed).

3.5 Restrictions following substantial road works

Background

3.5.1   The general public's image of street works is typified by a street being dug up within months of its being resurfaced whether by undertaker or by street authority. A primary objective of the co-ordination arrangements is to minimise the chances of this occurring in the future. Section 58 of the Act can assist in that process.

3.5.2   Under section 58 of the Act, the execution of street works may be restricted for a period of 12 months after the completion of "substantial road works". This provision has a two-fold purpose. Firstly, it aims to prevent undertakers breaking up streets within a short while of their having been resurfaced or reconstructed. Secondly, it seeks to avoid repetitive disruption of traffic by works being carried out in the street.

Substantial road works — meaning

3.5.3   The Regulations define this term; such works are street authority works either affecting the carriageway only, or the footway only, or both; cycle tracks are also included. They can consist of resurfacing, reconstruction, widening or alteration in the level of the part of the street concerned. The works must extend for more than 30 metres continuously. If in the footway, footpath, bridle-way or cycle track, they must extend over two-thirds of its width; if in the carriageway, they must affect more than one-third of the width. The 12-month restriction will apply only to the length of the street on which substantial road works have been carried out.

3.5.4   Specialist non-skid surface dressing falls within the meaning of 'substantial road works' but the street authority may decide not to exercise their powers under section 58 in all cases. Where this treatment is carried out without notice being given under section 58, the 12-month restriction will not apply.

Special requirements

3.5.5   The imposition of restrictions is conditional upon the street authority publishing a formal notice in a local newspaper and on their own website. Currently, a formal notice must also be published in the London Gazette at least 3 months prior to the works in question. Street authorities may in addition publish these proposed restrictions on their own website.

3.5.6　It is important that the proposed restriction is brought to the attention of all those who are likely to be affected by it, so that they may plan their works accordingly. Undertakers who have no immediate plans for street works in the street in question may wish to adjust their programmes in order to do their works in advance of the substantial road works, thus avoiding the delay which would otherwise ensue. It could enable the newly constructed or resurfaced street to remain undisturbed for much longer after completion. Copies of the notice must therefore be supplied to all undertakers and others likely to be affected (see paragraph 3.5.8(c) below).

3.5.7　Designations are recorded in the Additional Street Data sets of the NSG. For details see Chapter 4 Section 4.3.

3.5.8　Under the combined effect of the Act and the Regulations, the following provisions apply:

(a) the notice must be published in the prescribed manner and a recommended form is contained in Appendix C;

(b) it must specify the works, their nature and location, the date on which it is proposed to start them (not less than 3 months ahead), and clearly identify the extent of the restriction. There is no reason why the entire length of the highway should be subject to restriction if only a part of the length is to be affected by substantial road works;

(c) a copy of the notice must be given to all relevant authorities, to all undertakers who may have apparatus in the part of the street affected, to undertakers who have given advance notice under section 54 of the Act and to other persons who have asked to be supplied with copies of notices under section 58 of the Act. This notice may be given electronically in accordance with the data formats in Appendix E.

(d) the notice will be ineffective in imposing restrictions on street works unless the substantial road works in question commence within one month of the specified starting date, or if it is not possible to start them within that period because of street works in progress at the time, within one month of the completion of those street works.

3.5.9　Street authorities will, under section 53 of the Act, record all requisite details in their register.

Exemptions

3.5.10　Restrictions under section 58 of the Act do not affect the following:

— emergency or urgent works;

— works not involving breaking up the street, repairing, resetting and replacing manhole or chamber covers and frames, resurfacing up to 20 square metres, pole, lamp, column or sign replacement, pole testing, and works of a similar nature;

— works required to respond to a request for a new service or supply to a customer, which was not received at a time when it was practicable for the works to be done before the period of restriction began;

— other works to which the street authority have given their consent (which must not be unreasonably withheld).

Policy guidance

3.5.11   This power is an important one and street authorities are encouraged to use it in appropriate circumstances. Its use will not only protect streets where notice is served, but it should also encourage the development of a climate within which authorities and undertakers plan and amend their programmes in a way which is designed to minimise inconvenience for the public.

Undertakers should do their utmost to give details of their plans for works in affected streets as early as possible within the 3-month notice period and also to complete their works before the specified starting date. Nevertheless, if works within one of the exemptions (see paragraph 3.5.10 above) overrun, or have to be carried out after the specified date, it would be sensible from the point of view of road users for these street works to be accommodated before completion of the substantial road works concerned.

3.5.13   The exemptions described in paragraph 3.5.10 exemplify the balance that must be achieved if the co-ordination aspects of the legislation are to be successful. In considering applications for consent from undertakers, street authorities must take account of the needs of their customers. But equally, utilities must recognise the needs of road users and the need to ensure best value for money in highway expenditure. The key test is whether the undertaker could reasonably have foreseen the eventuality during the notice period and/or could reasonably be required to postpone the work until the end of the restriction.

3.6 Directions under section 56

Background

3.6.1   In the Act section 56 provides street authorities with an important power that is potentially onerous in terms of its impact on undertakers. Under this section an authority can serve a direction on an undertaker setting out times during which proposed street works may or may not be carried out. This power is subject to two important constraints:

— it can be used only where the proposed works are likely to cause serious traffic disruption that would be avoided or reduced if the works were to be carried out at specific times;

— an authority using this power must have regard to the advice set out in this Code of Practice.

3.6.2 It is not possible to define "serious disruption to traffic" in objective terms. However, in considering whether or not street works could cause serious disruption, and whether or not the timing of the works could have an impact, the street authority must have regard to:

— the nature, size and duration of the proposed street works;

— the normal level and speed of traffic on the street concerned;

— the nature of the traffic likely to be disrupted (whether, for example, the street is an important public transport corridor).

Cases where directions may be given

3.6.3 Directions may be given in the following circumstances:

(a) in a street which has been designated traffic-sensitive in order to curtail street works during traffic-sensitive periods, where the undertaker has not agreed to do so voluntarily;

(b) in a street not designated traffic-sensitive under ordinary conditions but which is required to carry traffic diverted from another street in order to accommodate street works in that street;

(c) where there is no traffic-sensitive designation but, by reason of a short-term local event or occurrence (e.g. a royal visit, road race, major sports event), it is necessary to curtail street works to avoid traffic disruption;

(d) where serious traffic disruption can be avoided or minimised by one work site serving two or more undertakers;

(e) where there is no traffic-sensitive designation but the scale of works is such that more than one traffic lane is affected and accordingly the works are likely to cause serious traffic disruption which can be avoided or reduced if the works are carried out only at certain times.

3.6.4 Directions cannot be given in the case of emergency works, nor in the special case of urgent works where the undertaker may proceed without advance notice (see paragraph 5.3.2(a)).

3.6.5 A direction should not be issued requiring street works to be executed out of normal hours if the effect would be to require the undertaker to commit a breach of any noise abatement or prevention provision under other legislation. Street authorities should therefore co-ordinate requirements with environmental health authorities before issuing a direction.

General considerations

3.6.6 It is important that section 56 directions are not used as an unsubstantiated veto on planned works.

3.6.7 It is also important that, if a direction is given, appropriate notice is given to the undertaker to enable him to adjust his working arrangements accordingly. The following arrangements have been agreed:

Undertaker's notice	Street authority's direction
2 hours	Within 1 hour of receipt
3 days	Within 1 day of receipt
7 days	Within 3 days of receipt
1 month	Within 10 days of receipt

3.6.8 A direction under section 56 may be given electronically in accordance with the data formats in Appendix E.

Policy guidance

3.6.9 Even in circumstances where the use of a section 56 direction may be appropriate a street authority should first endeavour to reach agreement with the undertakers concerned on the timing of works. The power of direction should be used only where:

— a voluntary agreement has not been reached; or

— the scale of works or potential for disruption is such that the street authority have reasonable grounds for seeking statutory backing for arrangements agreed in discussions with the undertakers; or

— there is insufficient time for agreement to be reached.

3.7 Section 85 Notices

3.7.1 Regulations made under section 85(2) of the Act allow an authority to give notice to undertakers of major highway, bridge or transport works where those undertakers' apparatus is affected by the works concerned.

3.7.2 The practical effect of such a notice is that the utilities will not receive a contribution to the cost of diverting any apparatus they choose to install after receiving the notice. This notice may be given electronically in accordance with the data formats in Appendix E.

3.7.3 Streets that have been subject to such a notice will be recorded in Additional Street Data in the NSG.

3.8 Other features of the street

Background

3.8.1 There are a number of other features of a street which may either:

— impact upon the planning and co-ordination of street works; or

— be subject to restrictions imposed by legislation other than the New Roads and Street Works Act 1991.

To facilitate best practice, information about such features may be held as Additional Street Data in the NSG. Data capture codes have been defined for the following types of feature:

Environmentally Sensitive Areas

3.8.2 These will include such areas as Sites of Special Scientific Interest and Ancient Monuments. The Special Designation Description will indicate the type of sensitive area.

Structures (not designated as being of Special Engineering Difficulty)

3.8.3 There are various structures associated with the highway which, whilst not fully meeting the criteria for designation as being of Special Engineering Difficulty, nevertheless warrant extra care being taken when working in their vicinity by following the guidelines given in Appendix G. The Special Designation Description will indicate the type of structure.

Special Surfaces

3.8.4 These include, but are not restricted to, such surfaces as Porous Asphalt, Tactile and Coloured Surfaces. The Special Designation Description will indicate the type of surface.

Pipelines

3.8.5 These are Government and Oil or Gas Pipelines laid under the Pipelines Act 1960.

Priority Lanes

3.8.6 These include Cycle Routes and Bus Lanes. The Special Designation Description will indicate the type of priority lane.

Level Crossing Precautionary Areas

3.8.7 The Special Designation Description will indicate the extent of the Precautionary Area. When works are proposed within the Precautionary Area, the extra precautions detailed in Appendix F of this Code must be applied.

Special Construction Needs

3.8.8 The Special Designation Description will indicate the extent and type of special construction and could include such sites as geotextile mats and areas where sulphate resistant concrete is required.

CHAPTER 4
The register

4.1 Background

4.1.1   Under section 53 of the Act each street authority is required to maintain a register with respect to each street for which they are responsible, containing information with respect to street works and other prescribed types of works. The effect of the Regulations is to extend the registration responsibility of highway authorities to private streets and to exempt street managers accordingly.

4.2 Contents

4.2.1   Regulations provide that the register may be kept in any form, so indexed as to enable relevant information relating to a particular street to be traced, and must record the following information:

   (a) Copies of all notices pursuant to sections 54, 55, 57 and 74 of the Act served upon the highway authority relating to street works in any street that is a maintainable highway

   (b) Copies of all notices pursuant to sections 54, 55 and 57 of the Act served upon street managers relating to street works in any street that is not a maintainable highway

   (c) Particulars of minor street works involving breaking up the street but not requiring prior notice in any street in the local highway authority's area.

   (d) Particulars of works for which plans and sections have been submitted under Schedule 4 to the Act (special engineering difficulties).

   (e) Particulars of notices given by any relevant authority under Schedule 4.

   (f) Particulars of street authority works for road purposes.

   (g) Particulars of street works licences.

   (h) Information as to completion of reinstatements provided under section 70(3) of the Act.

(i) The street gazetteer as supplied to the NSG concessionaire, capable of uniquely identifying each street.

The following items, as well as being recorded in the register, are available as Additional Street Data sets within the NSG:

(j) Details of all streets subject to special controls.

(k) Copies of notices published under section 58 of the Act.

(l) Copies of notices given under section 85(2) of the Act (disallowance of undertakers' costs).

(m) Details of road classifications for the purposes of undertakers' reinstatement obligations, and for use in determining the level of charges appropriate under section 74 of the Act (see Chapter 8).

4.3 Additional Street Data

4.3.1 When a highway authority make or withdraw a designation, or receives notification from a street manager of a designation, they must

(i) notify the concessionaire for the time being responsible for maintaining, updating and issuing the NSG; and

(ii) record such decisions on the Street Works Register on the next following quarter day, except where that day is a public holiday, in which case it shall must do so on the following working day.

"Quarter day" means 2 January, 1 April, 1 July, and 1 October.

4.3.2 For practical purposes, it is essential that the Additional Street Data available on the NSG remains in step with the Street Works Register. Therefore it is important that the Additional Street Data is forwarded to the concessionaire in sufficient time for the concessionaire to make it available on the NSG by the next quarter day. Details regarding sufficient time can be obtained from the NSG concessionaire.

4.4 Inspection of the Register

4.4.1 Everyone has a right to inspect the register, free of charge, at all reasonable times (i.e. normal office hours) except that the right is limited, in the case of "restricted information", to:

(a) persons authorised to execute any type of works in the street (ie. utilities, including cable operators with a franchise for the area, whether or not any of them have apparatus in the street, or have served notice of proposed works, and street works licensees), or

(b) persons "otherwise appearing to the authority to have a sufficient interest".

Any person claiming to see restricted information must satisfy the highway authority that his interest is greater than the general interest of the ordinary member of the public.

4.4.2 Restricted information is anything which has been certified by Government as being a matter of national security, or which the undertaker himself has certified as relating to a matter where his commercial interests may be jeopardised if the information were not restricted, e.g. a prospective contract under negotiation, or technical or scientific material.

4.4.3 Information provided by means of any notice under the Act should be retained for 6 years after completion of the guarantee period of the works referred to in the notice.

4.4.4 Highway authorities are encouraged to make their Registers available via publicly available Internet web sites.

CHAPTER 5

Work categories and notice requirements

5.1 Introduction

5.1.1 This Chapter describes the obligations in relation to the maintenance and the use of the street works register and the various notice and registration requirements of the Act and Regulations. It identifies the special requirements for emergency and urgent works and covers the full spectrum of work categories from very minor manhole inspections to cases where as much as one month's notice of starting date may be required.

5.1.2 Before an undertaker commences work for the first time in any particular highway authority's area that undertaker must contact that highway authority prior to serving his first notification in order to advise the highway authority of his DETR code and operational district etc.

5.1.3 The Act provides for different types of street works notice, including:

(a) notice of starting date — this must specify the proposed starting date and work must commence within a specified period. The Act mentions a 7-day notice and a 7-day period, but this can be, and has been, varied or dispensed with by Regulations in appropriate cases (section 55);

(b) advance notice — this only applies to certain prescribed cases, i.e. major projects and, in traffic-sensitive situations, both standard works and minor works involving excavations. The Regulations specify a one-month notice period (section 54);

(c) emergency works notice — this is given after beginning work. The Act specifies 2 hours, but this has been varied by the Regulations to deal with certain out-of-hours situations (section 57).

(d) the various new notices required under section 74 of the Act (see Chapter 8).

5.1.4 Figure 1 summarises the minimum notice requirements in relation to the various work categories. The definitions are contained in Appendix A. Appendix D explains the data that must be transmitted, together with the paper format, for non-electronic transmission, and Appendix E shows the data formats for electronic transmission.

5.1.5 The notice period for any work is determined by a combination of the nature of the work involved and whether or not it is to take place on a street subject to special controls. Before outlining the notice requirements in detail it is important to stress a number of general points.

5.1.6   The notice system performs at least five functions:

— it is a vital component of the co-ordination process; this is particularly important in the case of notices for works on traffic-sensitive streets and for major projects;

— emergency and urgent works can prompt the emergency procedures of other organisations;

— it triggers the inspection regime;

— it forms the basis of records for reinstatement guarantee purposes; and

— it facilitates the charging regime under section 74 of the Act.

5.1.7   The notice period starts when the recipient receives the notice, not when the notice is sent. In the absence of evidence to the contrary it is assumed that a notice sent by first-class mail will be received the following day and that a notice sent by fax, or electronic means, will be received on the same day. It is important that people completing and receiving notices understand these objectives and reflect them in the way in which they operate. It is particularly important that the information contained in notices is accurate and provides as much information as is available at the time at which the notice is completed. It is also essential that the notice periods referred to in this Code are treated as minimum periods. Wherever reasonably practicable longer notice should be given.

5.1.8   It is important to remember that, with the exception of the one-month advance notice of major projects and s74 notices which need only to be served on the street authority (see Chapter 8), any required notice must be served both on the street authority and on any other undertaker, or sewer, bridge or transport authority having apparatus in the street which might be affected by the works. However, in recognition of the impracticability of this requirement in some situations, the Regulations make provision for exemptions. Section 55 and 57 notices need not be served on undertakers unless an undertaker specifically requests it, nor to people who might otherwise be entitled to a notice solely by virtue of owning a drain or service pipe in the street.

5.1.9   In all cases where notices are required they will be entered in the street works register. There are cases where, although no notice is needed, works information will be registered; these are described in paragraph 5.4.6(b).

5.1.10 Provided everyone served with a notice agrees, works may start before the specified starting date.

5.1.11 In giving notice, undertakers should bear in mind the time required, in some cases, for formal road closure action (see Section 7.3 of this Code).

5.1.12 When works are to take place in a section of street designated as having special engineering difficulties, the undertaker must comply with the requirements of Schedule 4 to the Act, in addition to the appropriate notice requirements. Approval will normally be required in advance of issuing the notice and the requirements are fully described in Section 5.7.

5.1.13 It is a requirement of section 70(3) of the Act that undertakers must inform street authorities by the end of the next working day after they have completed reinstatement, and say whether it is interim or permanent. Registration of the reinstatement details should then be provided within 7 working days, using the data formats in Appendix E. See paragraph 7.7.2 for notice requirements when an interim reinstatement is replaced by a permanent reinstatement.

5.1.14 Each type of notice is valid for a specified period after which it will be deemed to lapse if work does not commence. The validity periods are described in Section 5.8. A notice may be cancelled if the undertaker knows that works are not to proceed in accordance with the notice; otherwise, where no works take place, the notice will lapse. As a notice will lapse in due course, there is no legal requirement to cancel it if the undertaker knows that works are not to proceed. However, it is good practice to issue a notification abandoning the proposed works rather than leave the notice of proposed works hanging, and undertakers are urged to do so.

5.1.15 If the proposed works are abandoned the Works_Promoters_Reference must not be used again for any other works, except in the case of Permanent Reinstatement Proposed and Remedial Reinstatement Proposed, when the original Promoter_Works_Ref for those works must be reused.

5.1.16 An undertaker's works may sometimes straddle the boundary between two or more street authorities. In such cases the relevant notices related to the works in their area should be sent to each authority in order, inter alia, to trigger inspection units and s74 charges should these latter become necessary. However, some authorities, not only street authorities but also other relevant authorities, may require to see notices of works in these adjacent areas, but in these cases action should be taken to ensure that neither inspection units are generated nor s74 charges are leviable. This can be accomplished by the use of Type 21 records within the Operational District Data Definition, see E3.3.1 and E3.4 in Appendix E. Nevertheless notices of works should not be sent to highway authorities except in relation to their own areas, unless requested via the creation of Type 21 records as an Other Interested in the SWA_Org_Type field.

Figure 1 Minimum Notice Periods		
CATEGORIES OF WORKS	Non traffic sensitive situations	Traffic sensitive situations
EMERGENCY (INCLUDING REMEDIAL — DANGEROUS)	WITHIN 2 HOURS OF WORK STARTING	
URGENT	WITHIN 2 HOURS OF WORK STARTING	2 HOURS NOTICE IN ADVANCE
SPECIAL CASES OF URGENT	WITHIN 2 HOURS OF WORK STARTING (WHERE IMMEDIATE START IS JUSTIFIED)	
MINOR WORKS (WITHOUT EXCAVATION)	NOTICE NOT REQUIRED	3 DAYS NOTICE
MINOR WORKS (WITH EXCAVATION)	NOTIFY BY DAILY WHEREABOUTS	ONE MONTH ADVANCE NOTICE AND 7 DAYS NOTICE OF START DATE
REMEDIAL WORKS (NON-DANGEROUS)	NOTIFY BY DAILY WHEREABOUTS	3 DAYS NOTICE
STANDARD WORKS	7 DAYS NOTICE	ONE MONTH ADVANCE NOTICE AND 7 DAYS NOTICE OF START DATE
MAJOR PROJECTS	ONE MONTH ADVANCE NOTICE AND 7 DAYS NOTICE OF START DATE	

Note: STREETS OF SPECIAL ENGINEERING DIFFICULTY & PROTECTED STREETS. Approval for works in such streets must be obtained from the Relevant Authority separately from, and in addition to, the formal notice.

5.2 Emergency works

Definition

5.2.1 Emergency works are defined in section 52 of the Act as those "whose execution at the time when they are executed is required in order to put an end to, or to prevent the occurrence of, circumstances then existing or imminent (or which the person responsible for the works believes on reasonable grounds to be existing or imminent) which are likely to cause danger to persons or property". The term also includes works not falling within that definition which cannot be severed from those that do, such as street works not at the emergency site that are necessary to shut off or divert a supply. Remedial works to dangerous defective reinstatements are emergency works (see paragraphs 7.7.6 to 7.7.8).

5.2.2 Bar holes used to detect gas leaks do not count as excavations and reinstatements for the purposes of registrations unless no other street works are carried out on the site when they should be registered under the normal rules.

5.2.3 The onus of proving the existence of an emergency always lies with the undertaker.

The notice requirement

5.2.4　In cases coming within the definition of emergency works, the undertaker may commence works without giving prior notice but must notify all concerned, under the requirements of section 57(2) of the Act, within 2 hours of works having begun.

Purpose of notice

5.2.5　The notice serves two purposes – firstly, to make the street authority aware of actual or potential disruption to traffic so that they can take appropriate action such as giving traffic information to the police and other interested parties, and secondly, to enable the excavation and reinstatement to be registered in accordance with the requirements of the Regulations.

Procedure

5.2.6　Where the emergency occurs during working hours, the undertaker will give notice within 2 hours of work commencing on site. Fax or electronic means may be used.

5.2.7　Where there has been a failure of fax or electronic transmission, three recorded attempts at fax or electronic transmission within the specified 2 hours will be seen as complying with the requirements. In this case, the undertaker must telephone the notice details to the street authority immediately and as soon as possible give them to other undertakers who have requested the information.

5.2.8　Where an emergency occurs outside normal working hours, and the street authority or undertaker have not established out-of-hours arrangements for notices, the undertaker will give formal notice within 2 hours of the start of the following working day, i.e. by no later than 10.00am.

5.2.9　Emergency works may be large jobs involving considerable disruption and the public expects as much warning as possible so that appropriate avoiding action may be taken. Undertakers should, wherever practicable, in addition to giving the formal notice, inform the street authority and if necessary the police by telephone of the emergency, as soon as the extent becomes apparent.

5.2.10　In the case of such large-scale emergency works street authorities should provide undertakers with details of officers who can be contacted by telephone, including those available for out-of-hours contact.

Traffic sensitive situations

5.2.11　The provisions relating to emergency works notices are the same for all streets, even in traffic-sensitive situations.

5.3 Urgent works

Definition

5.3.1 These are works which fall short of emergency works as defined in the Act, but are of sufficient urgency to warrant immediate action either to prevent further deterioration of an existing situation or to avoid an undertaker becoming in breach of a statutory obligation. They are defined in the Regulations as street works (not being emergency works) whose execution at the time they are executed is required (or which the person responsible for the works believes on reasonable grounds to be required):

— to prevent or put an end to an unplanned interruption of any supply or service provided by the undertaker;

— to avoid substantial loss to the undertaker in relation to an existing service; or

— to reconnect supplies or services where the undertaker would be under a civil or criminal liability if the reconnection is delayed until after the expiration of the normal notice period.

They include works that cannot reasonably be severed from such works.

5.3.2 Urgent work falls into two categories:

(a) those which must be acted upon without any delay even in traffic-sensitive situations where the same notice requirements for emergency works apply, i.e. no prior notice is needed, but notice to all concerned must be given within 2 hours at the most. These are "special cases of urgent works" (see paragraph 5.3.3 below); and

(b) the remainder, for which 2 hours prior notice is required in traffic-sensitive situations to allow time for local traffic management arrangements to be made. These are "other urgent works" (see paragraph 5.3.9 below).

Special cases of urgent works

5.3.3 Special cases of urgent works are those which are required, or are believed to be necessary, for restoring or preventing an interruption in a supply or service to premises where:

(a) the undertaker and the street authority have agreed in advance that immediate action may be taken in the event of a loss of service occurring or threatening (e.g. hospitals, old people's homes, stock exchange or dealers' premises, banks, police stations, premises with telephone alarm systems, commercial and industrial premises reliant upon continuous supplies etc.); or

(b) the undertaker fears that an emergency situation may develop if immediate action is not taken and considers that the need for such action outweighs the danger of traffic disruption.

Procedure

5.3.4 The procedure and position regarding traffic-sensitive situations are the same as in paragraph 5.2.10 above.

5.3.5 In addition, the undertaker will, as soon as practicable, inform the street authority in writing of the grounds upon which he considered immediate action to be necessary, if the information has not already been given in the Works Description Text of the original notice.

Policy guidance

5.3.6 As works may be undertaken under this provision without prior notice, even in designated traffic-sensitive streets at traffic-sensitive times (i.e. traffic-sensitive situations), it is of the greatest importance that undertakers satisfy themselves that any works undertaken in such cases without the consent of the street authority can be fully justified. Any disagreement between the undertaker and the authority should be referred in the first instance to the conciliation and arbitration machinery described in Chapter 9.

5.3.7 It is essential that undertakers use this provision responsibly. It is equally important that street authorities recognise the needs of utility customers and the impact that a loss of supply can have on key institutions and organisations. This provision will prove particularly useful where an undertaker fears that persons or property may be endangered if immediate action is not taken, e.g. in the reconnection of a group of houses rather than an individual property.

5.3.8 In no circumstances may new supplies or services be installed without notice in a traffic-sensitive situation, unless prior agreement of the street authority is obtained.

Other urgent works

5.3.9 These are other cases of works which are not emergencies, but which are required (or believed on reasonable grounds by the person responsible for them to be required):

— to deal with an actual or threatened interruption in any supply or service provided by the undertaker, or

— to deal with substantial loss to the undertaker in relation to an existing service, or

— to restore a supply or service that has been cut off, where the undertaker would be under a civil or criminal liability if reconnection is delayed until after a normal notice period has expired.

The notice requirements

5.3.10 This and the following three sections cover all cases of urgent works except the special cases described in paragraph 5.3.3. The notice requirement differs between works in a traffic-sensitive situation and other cases.

5.3.11 **Traffic-sensitive situations.** Two hours prior notice must be given. Work may only start before expiry of the notice with the agreement of the street authority.

5.3.12 Where a situation justifying urgent action occurs outside normal working hours, and the street authority or undertaker have not established out-of-hours arrangements for notices, the undertaker will give formal notice within 2 hours of the start of the following working day, i.e. by no later than 10.00am.

5.3.13 **Non-traffic-sensitive situations.** As there is little risk of the works causing serious traffic disruption, and because of the urgency of the need to proceed with the works as soon as practicable, no notice is required before their commencement. The undertaker must however give notice within 2 hours of commencing. The procedural requirements in such a case are similar to those described in Section 5.2 relating to emergency works.

Purpose of notice

5.3.14 The purpose of a notice of urgent works, whether prior notice, or notice given after commencement, is similar to that of the emergency works notice (paragraph 5.2.4).

5.4 Minor works

Definition

5.4.1 To qualify as minor works the works must:

(a) not be emergency or urgent works, and

(b) not be of a planned duration of more than 3 days, and

(c) not form part of a rolling programme, and

(d) not be planned to involve at any one time more than 30 metres of works or leave less than the minimum width of carriageway necessary for one-way traffic as defined in the Code of Practice *Safety at Street Works and Road Works*.

They may be carried out in the footway, verge or carriageway; they can range from small operations in existing manholes to the provision of a new supply to a customer.

5.4.2 Different notice requirements will apply, depending on a combination of two factors:

(a) whether excavation is involved, and

(b) whether work takes place in a traffic-sensitive situation.

Minor works not involving excavation

5.4.3 There is no definition of this class of minor works but typically they include:

— works at manholes and chambers;

— replacement of poles, lamps, columns and signs not involving any change in siting;

— pole testing;

— operating valves, or

— works of a similar nature.

5.4.4 The notice requirements for such works are as follows:

(a) for works to be undertaken in a traffic-sensitive situation — 3 days advance notice;

(b) otherwise — because they have no significant effect on traffic, no notice or registration is required.

Minor works involving excavation

5.4.5 These are minor works, as defined above, which involve some degree of excavation, such as:

— laying or repairing service pipes and cables;

— joint repairs to pipes and cables not in manholes;

— valve replacements;

— resurfacing up to 20 square metres;

— other works to remedy defective reinstatements not constituting a danger to the public; and

— relatively small excavations such as resetting frames and covers.

5.4.6 The following notice requirements apply to these works:

(a) if proposed to be carried out in a traffic-sensitive situation — advance notice of one month followed by 7 days notice of starting date;

(b) otherwise — no prior notice is required, but the street authority must be informed by the start of the works by means of the "Daily Whereabouts" using the data formats shown in Appendix E for electronic notices, or the forms in Appendix D for non-electronic notices to enable the works to be included in a sample for inspection.

(c) for the avoidance of doubt, if a works is not started on the day stated on the "Daily Whereabouts", then a new "Daily Whereabouts" must be issued before the works can start.

5.4.7 The need for compliance with section 70(3) of the Act must not be overlooked (see paragraph 5.1.13).

Purpose of the notice

5.4.8 The 3-day notice (paragraph 5.4.4(a) above) provides an opportunity for the street authority to influence the hours of work, if necessary, by a direction under section 56 of the Act, where traffic considerations make this necessary.

5.4.9 Where, because the danger of traffic disruption is greater, one month's notice is needed (paragraph 5.4.6 above), the street authority have an opportunity, by discussion with the undertaker and other relevant parties, to co-ordinate the works with other proposed activities in the street, and also to make any traffic management arrangements which the circumstances require.

Procedure

5.4.10 The appropriate notice may be served by fax or electronic means, post or hand delivery, provided it complies with the period of notice requirements.

5.4.11 The response times for those receiving a notice who wish to discuss any amendment to the proposals are, for a 3-day notice, 1 day, and, for a 1-month notice, 10 days, in each case from the date of receipt of the notice.

5.5 Standard works

Definition

5.5.1 These are works which are not:

— emergency works (paragraph 5.2.1)

— urgent works (paragraph 5.3.1), or

— minor works (paragraph 5.4.1).

They include resurfacing works in excess of 20 square metres and cases where an undertaker returns to site after interim reinstatement to complete permanent reinstatement in excess of 20 square metres.

The notice requirements

5.5.2   (a) Works in a traffic-sensitive situation — advance notice of at least one month and 7 days notice of starting date.

(b) Other works — 7 days notice of starting date.

Purpose of the notice

5.5.3   The aims of the longer notice periods for standard works are:

— to facilitate co-ordination and ensure minimum disruption to traffic on traffic-sensitive streets;

— to provide time for negotiations between undertakers and the street authority in relation to work on protected streets.

5.5.4   Standard works, by definition, may be of some significance in terms of size, duration and impact upon the street authority, road users and other undertakers, whenever they are carried out. The requirement to provide advance notice in traffic-sensitive situations is an essential tool for satisfactory planning, management and co-ordination of works by the street authority. One month is the minimum notice period, and where possible undertakers should endeavour to give more notice than this.

5.5.5   The street authority is generally the first point of contact for public complaints. In order to ensure the public is made aware who is responsible for the works, it is important that the promoting undertaker gives advance information and warning to affected frontagers not only about any disruption to their services, but also if their access is to be affected for any length of time. Details of alternative access arrangements, or any other form of mitigating action to be taken by the promoting undertaker, should be supplied. (See also paragraphs 3.4.16 to 3.4.18.) If the undertaker produces any special publicity he should copy it to the highway authority.

Procedure

5.5.6   The appropriate notice may be served by fax or electronic means, post or delivery, provided it complies with the period of notice requirements.

5.5.7   The response times for those receiving a notice who wish to suggest any amendment to the proposals are, for a 7-day notice, 3 days, and for a 1-month notice 10 days, in each case from the date of receipt of the notice.

5.6 Major projects

Definition

5.6.1 Major projects are "standard works" which have been identified specifically in the undertaker's annual operating programme or which, if not specifically identified in that programme, are normally planned at least 6 months in advance of works commencing.

Purpose of notice

5.6.2 Undertakers are required under the Regulations to give one month's advance notice of major projects.

5.6.3 This notice is primarily concerned with the objectives of co-ordination described in Chapter 6. It provides an opportunity for the street authority to consider the implications of all proposals for street works which are likely to have a major impact upon traffic, the structure of the highway or the apparatus or proposals of other undertakers, possibly giving rise to the need for road closures and diversions, and where there is greater need to co-ordinate the proposed works with those of other undertakers and the street authority themselves.

5.6.4 In the case of major projects, by definition, the undertaker will, in formulating his proposals, have made them known to, and discussed them and their programming with, the street authority and other interested parties. Although the formal legal requirement is for one month's notice of such projects, undertakers must always use their best endeavours to give at least three months' advance notification of proposals for extensive works together with route plans. Street authorities for their part must give the same advance information of their own major highway works. Provision of advance information will reduce the likelihood of programme changes being required at a later date.

5.6.5 This procedure will enable proposals from both sides to be discussed at the regular co-ordination group meetings described in Chapter 6. They may even give rise to the need for a special meeting of the group.

5.6.6 Both street authorities and undertakers, in cases where diversionary works are involved, will have to comply with the requirements of the Code of Practice *Measures Necessary where Apparatus is affected by Major Works (Diversionary Works)* under section 84 of the Act relating to such works. If proper planning and costing of works is to occur, it is desirable that longer periods of advance notification should be given than even the three months minimum required by section 58(2) of the Act (*Restriction on works following substantial road works*).

5.6.7 In some cases where the undertaker considers that confidential factors affecting either national security or commercial confidentiality are involved, he may not feel able to give more than the minimum notice of his proposals. He should, however, consider taking the street authority into his confidence in these circumstances.

Procedure

5.6.8 Service of the notice will be by fax or electronic means, post or hand delivery. A one-month notice should be responded to within 10 days if alternative requirements are proposed.

5.7 Works in streets subject to special controls

5.7.1 Care must be taken to observe the special requirements that apply to works in the streets, which are subject to special controls (described in Sections 3.2, 3.3, 3.4 and 3.5). These are:

— protected streets (sections 61 and 62 of the Act)

— streets with special engineering difficulties (section 63 of the Act)

— traffic-sensitive streets (section 64 of the Act)

— streets where substantial road works have been carried out (section 58 of the Act)

PROTECTED STREETS

5.7.2 Undertakers' works in verges and central reservations not impinging on the carriageway should normally be allowable during the day, but other street works, wherever practicable, will only be undertaken at night or, where appropriate, at weekends, or otherwise at times when the impact upon traffic will be minimised.

5.7.3 Undertakers should give the same notice for protected streets as those applicable to traffic-sensitive streets.

Consent

5.7.4 In the case of works involving the placing of utility apparatus in a protected street (not being renewal of existing apparatus) there is an additional need to obtain the consent of the street authority. Conditions may be attached to the consent, and in so far as they involve the undertaker in additional expense, the street authority may contribute to the cost of compliance.

5.7.5 Protected streets are not intended to be a barrier to the extension of undertakers' networks, and accordingly, in the absence of special reasons, undertakers may expect consent to be given to works crossing the protected street. Similarly, undertakers' works in verges and central reservations should normally be acceptable if they do not impinge on the carriageway.

5.7.6　Street authorities may charge a reasonable fee, initially and annually, to cover expenses incurred by them in relation to consents.

STREETS WITH SPECIAL ENGINEERING DIFFICULTIES

5.7.7　The special feature of these streets is that an undertaker proposing to carry out street works must not only give prior notice, and, if appropriate, observe the special requirements for traffic-sensitive streets (see the earlier sections of this Chapter), but must also comply with the requirements of Schedule 4 to the Act. Under that Schedule, plans and sections or, in appropriate cases, a description of the works, have to be submitted to the relevant authorities and, except in the case of emergency works, settled in advance of works starting.

5.7.8　In this context, the relevant authorities are the street authority who made the designation and the transport, bridge or sewer authority whose structure the designation is intended to protect. The term does not include an undertaker who might have an interest in the designation (but see paragraph 5.7.14 below).

5.7.9　The following paragraphs provide practical guidance to undertakers and also street authorities and other relevant authorities on the operation of the procedures in relation to these streets, including, in particular, the circumstances in which a plan and section drawing may be dispensed with.

5.7.10　Undertakers are encouraged to discuss their proposed works with the relevant authority as early as possible. Early discussions concerning the scope and nature of the proposed works may enable the authority to waive their requirement for a plan and section drawing where other arrangements prove adequate.

Emergency and urgent works

5.7.11　In these cases, the relevant authority must be notified as soon as reasonably practicable after the commencement of the works, i.e. in practice within 2 hours, as for the emergency works notice to the street authority. At this stage a description of the works will be sufficient. It is essential that relevant authorities provide undertakers with emergency contacts for this purpose, as early notification of these works can often provide an opportunity for the parties concerned to discuss the approach to be adopted in dealing with the emergency.

Works not requiring plans and sections

5.7.12　Street works that can be undertaken either without breaking up or opening the street, or by making use of existing manholes, cannot endanger the integrity of the street or structure that the designation was intended to protect. Accordingly, the following principles apply:

(a) No action under Schedule 4 is required in respect of any of the following works: -

— those not involving any breaking up or opening of the street;
— small works in manholes and surface boxes, such as cable repairs, meter replacements and valve operations;
— pole, lamp, column and sign replacement in situ;

- pole testing;
  - resetting frames and covers;
  - resurfacing areas up to 20 square metres;
  - remedial works to restore reinstatement to specification.

(b) Works involving the insertion or extraction of cables or ducts from existing manholes should be notified under Schedule 4, but a description of the works will be sufficient and the relevant authorities' response, if any, should be made within 3 days.

(c) Works to a service pipe or line in a bridge where the apparatus is in preformed pipe or cable bays, and can be reached by removal of the paving or loose fill, will also require only a description of the works; in this case, however, the relevant authorities will have 7 days within which to respond.

Works requiring plans and sections

5.7.13 The following principles apply where works involve breaking up a street and require plans and sections:

(a) To facilitate interpretation of plans and sections, details of the civil engineering design and requirements and method of construction and implementation of the works should, where appropriate, be submitted to the relevant authority. It is, therefore, essential that the authority is informed of proposed works as far in advance as possible, to allow the effect on the structure to be determined. This should be accomplished by informal exploratory discussion in addition to co-ordination meetings as described in Chapter 6.

(b) When exploratory discussions show that little if any alteration is required to the structure concerned, a plan and section should be submitted to the relevant authority. If the works relate only to a service pipe or service line, in each case of less than 100 metres in length, or overhead electric or telecommunications lines, the submission need allow only 7 working days for response. Otherwise the submission must allow one month for approval or disapproval.

(c) Where it is clear that more than minimal alteration to the structure will be needed, plans and sections should be submitted in good time to allow for adequate consideration of the proposed works and the protective measures or alterations that must be made to the structure concerned, and the appropriate response (approval or otherwise) from the relevant authority.

5.7.14 Where a street has been designated because of undertaker's apparatus or a hazardous pipeline which is fundamental to the structure and integrity of the street or is particularly sensitive to the risk of damage by street works (see paragraph 3.3.6(h)), the street authority shall must consult that undertaker when plans and sections of proposed works are submitted. The authority must not approve any proposals except in accordance with the specified requirements of that undertaker, and, in cases of dispute between undertakers, the plans and sections will have to be settled by arbitration in accordance with Schedule 4 to the Act.

5.7.15 The undertaker promoting the works should afford the relevant authority all reasonable facilities for inspecting or monitoring the execution of the works. The extent of the support needed will depend on the scale and nature of the works.

5.7.16 In the case of minor works, an inspection to ensure that the structure remains undamaged will be sufficient.

5.7.17 Extensive works, or any underground works immediately adjacent to the structure concerned, can involve monitoring the excavation, installation and maintenance of supporting works or structures and the backfilling of excavations.

5.7.18 In lieu of inspection or monitoring, the relevant authority may accept from the undertaker a certificate (or stage certificates) of compliance with the approved plans and sections and working methods.

5.7.19 The undertaker promoting the works will be responsible for the reasonable costs of taking measures to protect the relevant structure.

5.7.20 Where the relevant authority is legally entitled to recover his reasonable costs and it is necessary to monitor the works, the undertaker must meet the reasonable monitoring costs. Should the undertaker dispute the need for monitoring in whole or in part, to save delays to the works the monitoring must take place and be charged to the undertaker, and the dispute must be settled subsequently.

TRAFFIC-SENSITIVE STREETS

Notice period

5.7.21 The minimum standard notice period for works proposed to be carried out in a traffic-sensitive situation is one month, with a firm start date being notified at least 7 days in advance.

5.7.22 The only exceptions to the one-month notice requirement are:

(a) emergency works and special cases of urgent works — notice within 2 hours (paragraphs 5.2.4 and 5.3.2(a));

(b) other urgent works — 2 hours' prior notice (paragraph 5.3.11);

(c) minor works where no excavation is involved — 3 days' notice (paragraph 5.4.4);

(d) remedial works to bring failed reinstatements back to specification — 3 days' notice (paragraph 7.7.8 and Figure 2); and

(e) works which are closed down during traffic-sensitive times so as to allow unimpeded traffic flow.

Special requirements

5.7.23 These special requirements apply both to:

(a) undertakers carrying out street works, and

(b) highway authorities carrying out works for road purposes.

5.7.24 Planned works that will be affected by a designation need careful co-ordination to ensure that:

(a) they are organised so that, subject to undertakers' obligations under section 65 of the Act, the amount of road space devoted to working areas is kept to the minimum (this can involve plating excavations where it is safe to do so, when works are not actually in progress);

(b) works of the undertakers and the highway authority take place together, or as part of a continuous programme, to avoid repeated disturbance of the same section of the street;

(c) any necessary additional discussions are held between the street authority, utilities, police, bus companies, emergency services, etc., regarding suitable diversion routes so that they do not clash with other works;

(d) in the case of major works, the public are given appropriate advance information of the works by means of signs advertising their purpose, proposed starting date and duration, and recommended diversions, and also by press releases for media publication;

(e) the notified time of start and duration of the works is adhered to;

(f) consideration is given to the use of trenchless techniques where practicable;

(g) if works are coned off and a traffic lane is blocked but no works appear to be in progress, there should be an explanation of why the blockage of the traffic lane is still necessary;

(h) consideration is given to the hours of working each day, with due regard to local residents, to reduce overall traffic disruption;

(i) due regard is given to the possible need for formal road closures.

5.7.25 Undertakers should clearly understand that shorter notice periods for works in traffic-sensitive streets are only permissible for works that take place outside sensitive times. Where street works are planned to take place in a traffic-sensitive street outside traffic-sensitive times and, as a consequence, a shorter notice period, or no notice, is given, it is essential that the works are suspended during the sensitive period. Full, unimpeded traffic flow must be provided during the sensitive period, if necessary by temporarily backfilling excavations or using road plates.

5.8 Notice Validity

5.8.1 The one-month's advance notice for work in traffic-sensitive situations must specify a provisional starting date. At least 7 days before that date the undertaker must give a 7-day notice confirming the starting date. The starting date must be not later than one month from the provisional date given in the one-month advance notice.

5.8.2   Unless it is otherwise agreed, a 7-day notice shall ceases to have effect if work is not started within 7 days of the starting date specified; for a 3-day notice, the relevant period is 3 days. Outside those periods, a new 7-day or 3-day notice must be served, referring to the original notice.

5.8.3   The undertaker's prior discussions with all those likely to be affected by his proposals should, in almost every case, have provided the opportunity for any necessary modifications to be made prior to the formal one-month notice. However, should there remain a need for the works specified in the notice, or their timing or manner of execution, to be modified at this stage at the instance of the street authority or other authority, the authority concerned must notify the originator within 10 days of service of the notice. The street authority must be kept informed of any modifications that are made.

5.8.4   One month's advance notice for major works must specify a provisional starting date and must be followed by a 7-day notice (or notices) specifying the intended starting date (or dates) when known.

5.9 Restrictions following substantial road works

Notice requirement

5.9.1   The length of notification which an undertaker should give for works which he wishes to carry out during the period of a restriction imposed under section 58 of the Act depends upon whether:

(a) the street works come within the scope of any of the specific exemptions (e.g. emergency, urgent or minor works, or works unavoidably required to provide a new service to a customer), or

(b) the street authority's consent is required.

5.9.2   In case (a) the ordinary rules appropriate to the works concerned, set out in earlier Sections of this Chapter, must be followed.

5.9.3   In case (b), in addition to compliance with (a) above, application for consent should be made at least one month in advance, specifying, in addition to the normal works information, the grounds upon which consent is sought.

5.9.4   The provisions enabling the street authority to give consent to works cover other unforeseen cases not within the scope of the specific exemptions, which it is difficult to define in advance, e.g. diversions of apparatus to accommodate works elsewhere and replacement of lead water service pipes that fail relevant water quality requirements.

5.10 Street authority's works for road purposes

5.10.1 The Regulations require registration of works for road purposes as well as street works. Such registration identifies responsibility for the works; it also reflects the importance of advance notification of works of all kinds in the street if they are to be effectively planned and co-ordinated, to minimise disturbance of the road structure and limit disruption to road users.

5.10.2 Accordingly, street authorities, as part of their co-ordination function, will register details of the following works:

(a) works in protected streets, streets with special engineering difficulties and traffic-sensitive streets; major highway projects specified in the authority's annual programme or which are planned to commence at least six months in advance

        ONE-MONTH REGISTRATION

(b) other works   **7-DAY REGISTRATION**

5.10.3 Registration is not required for pothole repairs, road markings, resurfacing not exceeding 20 square metres in area, replacing and resetting slabs and similar works.

5.11 Noticing Procedure — Trench Sharing

5.11.1 Because trench sharing can minimise street disruption, both highway authorities and undertakers wish to encourage such working practices. However, it must be stressed that there can be no imposition of such methods of working.

5.11.2 The emphasis in such circumstances must therefore, be on mutual co-operation between all interested parties in order to derive the obvious benefits for the travelling public and utility customers.

5.11.3 In the event of trench sharing one undertaker (the <u>primary</u> undertaker) should take overall responsibility as the agreed point of contact with the highway authority. The other one (or more) secondary undertaker(s) will retain the same responsibility for submitting notices in accordance with the Act indicating the work carried out by them or on their behalf.

5.11.4 Only those notices submitted by the primary undertaker are required to show the estimated inspection units attributable to the works. The primary undertaker must, in the initial notice, detail the other undertakers involved and the scope of the trench sharing in the Works Description Text. All other undertakers should submit the correct notices on

which must be clearly marked, within the Works Description Text, that "trench sharing is involved". The secondary undertakers's notice will also indicate that their works are without excavation and therefore will not contain inspection units. The primary undertaker must also ensure that estimates of works duration are agreed and/or confirmed with the secondary undertaker(s) when submitting notices to comply with section 74 charging requirements.

5.11.5 By local agreement it should be possible to contractually arrange that the excavating primary undertaker serves notice and carries out work on behalf of itself and others. However, it must be emphasised that such arrangements cannot remove legal liability imposed by the Act on individual undertakers.

5.12 Other Authorities

5.12.1 Other authorities means other relevant authorities as mentioned in section 55 of the Act together with other interested authorities (ie highway authorities which have an interest in a street within, or immediately adjacent to their own geographic boundary, but which are not the street authority for that street).

5.12.2 Notices of proposed works (ie notices under sections 54, 55, 57 and Daily Whereabouts), notice of Actual Start Date, notice of Works Clear and notice of Works Closed should be copied to other authorities.

CHAPTER 6
Co-ordination in action

6.1 Information — the key to co-ordination

6.1.1 To ensure successful co-ordination there must be a free flow of accurate information and good communication between street authorities and utilities. The street authority is responsible under section 59 of the Act for the co-ordination of proposed works affecting the highway, whilst utilities are required to co-operate in this process under section 60 of the Act. This co-ordination can only be achieved if all relevant bodies adequately inform the street authority by notice of all proposed works, except those with or likely to have no or minimal impact on others.

6.1.2 Where the need for co-ordination is the greatest, longer notice periods than the basic seven working days are desirable. Notice of seven days or less, whilst necessary from the viewpoint of inspection and establishing responsibility for the works, gives little opportunity for co-ordination (which, because of the type of works involved, will not generally be called for).

6.1.3 Where the need for co-ordination arises, co-operation will be the keynote to effectiveness. This means that whoever proposes works in a street must be prepared to give timely advance notification of them, to discuss them with the other interested parties, including frontagers, and where appropriate and practicable modify them.

6.1.4 All parties must recognise the advantages of giving more advance information and notification than the prescribed minimum notice. But there will be circumstances where the street authority and others concerned may have no objection (or indeed, find it advantageous) if an undertaker proceeds before the end of the full prescribed notice period. In such cases they should give their consent accordingly.

6.1.5 In the case of streets subject to special controls, notice longer than the basic statutory seven days will be required, to allow time for street authorities, in conjunction with others concerned, to co-ordinate works.

6.1.6 The National Street Gazetteer and Additional Street Data provides basic details of streets, restrictions and special characteristics.

6.2 Co-ordination machinery

6.2.1 Co-ordination means resolving any differences between those competing for space or time in the highway, including traffic, in a positive and constructive way. In doing so, the statutory service obligations of undertakers and the statutory objectives contained in sections 59, 60 and 66 of the Act should be borne in mind.

6.2.2 Thus, street authorities and undertakers need to be given early warning of each others' major projects, so that when the project arrangements are finalised, these take into account the former's requirements and those of any related projects. This will entail assessing the likely problems created by works proposals in the light of:

(a) the road network capacity at the relevant times;

(b) the scope for conjunctive working arrangements, where appropriate and practical, between different undertakers and the street authority;

(c) the optimum timing of the works from all aspects;

(d) their effect on traffic and, in particular, the need for temporary traffic restrictions or prohibitions;

(e) discussion of appropriate techniques and arrangements particularly at difficult road junctions and pinch points;

(f) the special working arrangements required in streets with special engineering difficulties.

6.2.3 In many cases the street authority will be able to co-ordinate works effectively on a one-to-one basis with the other party concerned. However, co-ordination will for the most part be achieved through regular meetings of dedicated groups comprising representatives from all appropriate major interests, including not only the highway authority and undertakers, but, as the occasion requires, the local planning authority, police, other emergency services and organisations representing disabled people.

Regional HAUCs

6.2.4 At regional level the groups will be set up under the aegis of regional HAUCs and will be concerned principally with policy determination within national HAUC guidelines, monitoring the effectiveness of local co-ordination arrangements, providing policy guidance on a local basis and, where possible, resolving local differences. They will also facilitate conciliation and arbitration procedures. If the parties wish, performance reviews may also be carried out at these meetings.

Local Co-ordination

6.2.5 At local level these groups should be organised and chaired by the relevant street authority. They may be convened at an area level (e.g. County level) where appropriate, but wherever possible the groups should be based on a highway authority managed area and encompass such smaller street authorities as may be relevant in order to minimise the number of meetings. They will be concerned primarily with direct co-ordination of individual schemes and dissemination of information.

6.2.6 These local groups should meet quarterly but may meet at other intervals if the street authority and utilities so agree. They should cover the following topics:

— specific major projects, medium term and annual works programmes for both highway authorities and undertakers, submitted at least 21 days prior to the meeting;

— local policies and strategies affecting street works, traffic management proposals (including the effect of diversionary routes), and the potential for reducing disruption through common schemes/trench sharing etc;

— proposed designations of streets subject to special controls and other constraints;

— reviews of performance at local level, including damage prevention;

— feedback from HAUC;

— street works licences.

6.2.7 Representatives from all major interests should ensure that they are well enough informed to be able to discuss major projects, and medium-term and annual work programmes that are relevant to them. They should be able to demonstrate the appropriate knowledge of individual schemes where these are of concern to their organisations, and should also be able to speak and take appropriate decisions on behalf of their organisations.

Terms of Reference

6.2.8 Model terms of reference for both Regional HAUCs and local co-ordination meetings have been established by national HAUC, and can be found in Appendix H.

6.3 Licensees

6.3.1 Street authorities will maintain a record of all works and apparatus installed under street works licences granted by them. Street authorities, in responding to a plant enquiry regarding their own apparatus, should also include details of any licensees' apparatus

6.4 Liaison with other bodies

6.4.1   Street authorities should liaise with other bodies which have an interest in street works, and establish channels of communication to provide information to such agencies as the following:

- adjacent street authorities;

- the police;

- other emergency services;

- public transport operators; and

- any other appropriate bodies, e.g. organisations representing disabled people, pedestrians, motorists, and cyclists.

CHAPTER 7
Related matters

7.1 Location of works and apparatus — exchange of information

7.1.1　It is important that adequate information is provided to all concerned about the location and nature of all relevant apparatus. Section 79 of the Act, which is not yet in force at the time of going to print, makes various provisions for the recording of the location of apparatus. The detailed requirements will be set out in the code of practice on recording of underground apparatus when it is published.

7.1.2　Until there is the facility for all interested parties to exchange information electronically, existing arrangements should continue to ensure that information about apparatus is made available to undertakers. Street authorities, in responding to a plant enquiry regarding their apparatus must also include details of any licensee's apparatus. This is important in order to avoid damage to underground apparatus, and to comply with the Health and Safety Executive's requirements (current publication HS (G) 47 "Avoiding danger from underground services").

7.2 Prospectively maintainable highways

7.2.1　Section 87 of the Act provides that, where a highway authority are satisfied that a street in their area is likely to become a maintainable highway, they may make a declaration to that effect and such a declaration must be registered as a local land charge. Such a street must also be included in the local street gazetteer, which, amongst other things (see Section 4.2), will be expected to identify the following:

(a)　publicly maintainable highways;

(b)　prospectively maintainable highways;

(c)　private streets that are public highways of which the highway authority has knowledge, together with details of the street manager, where that is known.

7.2.2 This Code applies to prospectively maintainable highways just as it does to publicly maintainable highways.

7.2.3 In the case of private streets that are public highways, notices should be served on the street manager. A copy should also be supplied to the highway authority for registration purposes.

7.2.4 Private streets and prospectively maintainable highways are to be identified in the Additional Street Data set of the NSG. See Appendix E.

7.3 Road closures and traffic restrictions

Background

7.3.1 Sections 14-16 of the Road Traffic Regulation Act 1984, as amended by the Road Traffic (Temporary Restrictions) Act 1991, and Regulations which have been made under the 1984 Act prescribing the procedures, contain the relevant provisions governing temporary road closures and traffic restrictions for street works. There are two separate procedures, depending upon the urgency of the works.

7.3.2 Where the traffic authority are satisfied that urgent action is needed, they may issue a "temporary notice" imposing a closure or restriction; no prior notice need be given, but only a short-term closure or restriction is possible.

7.3.3 If there is a risk of danger to the public or serious damage to the road independent of street works (e.g. a leaking gas main) the notice is limited to 21 days duration, which can be extended by one further notice. In cases not involving a risk of danger or damage, the notice may only last 5 days.

7.3.4 In less urgent cases to which the preceding paragraphs do not apply, the traffic authority may make a "temporary order", which in the majority of cases may remain in force for up to 18 months (but only 6 months in the case of footpaths, bridleways, cycle tracks and byways open to all traffic).

7.3.5 Both a temporary notice and a temporary order may provide that restrictions have effect only when traffic signs are lawfully in place. In cases where street works are progressive along a length of road, this will assist in limiting traffic disruption.

7.3.6 The Road Traffic Act 1991 allows for the temporary suspension by the police of designated street parking places in order to prevent or mitigate traffic disruption or danger to traffic in exceptional circumstances. This could prove useful to undertakers at times when they are carrying out emergency works.

Procedure

Temporary notices

7.3.7 This procedure will generally apply to unplanned works (whatever the size of the job), e.g. emergency works, as well as urgent works responding to dangerous situations (see paragraph 7.3.3 above); it may also be appropriate for minor works involving an element of urgency not associated with dangerous situations.

7.3.8 In all the cases referred to above the undertaker will inform the traffic authority as soon as practicable if a closure or traffic restriction is needed. The authority will, as a matter of practice, consult with the police and all relevant parties and confirm as soon as possible whether or not a notice will be made.

7.3.9 The traffic authority is required in the notice to state the reason for which it is issued, its effect, alternative routes available (where applicable) and the date and duration of the notice. In addition, the traffic authority must give notice, on or before the day the notice is issued, to the police, to the fire authority, and to any other traffic authority whose roads are likely to be affected.

Temporary orders

7.3.10 The traffic authority must publish notice of intention to make a temporary order at least 7 days in advance. Where it is proposed to make an order for works expected to last more than 18 months, the proposal must be advertised at least 21 days in advance and published, not only in the local newspaper, but also in the London Gazette. No time limit applies to such orders, but they must be revoked as soon as the works are completed.

7.3.11 On or before the day on which a temporary order is made, the authority must notify the police, the fire authority (where appropriate), and any other traffic authority whose roads are likely to be affected. In the case of closures intended to last more than 18 months, the authority must not only notify but also consult these bodies.

7.3.12 Except in the case of an order following a closure notice (see paragraph 7.3.2 above), the need for an order of this type will generally occur in the case of planned works of which the undertaker gives at least 7 days notice. However, because of the time taken to consult, to obtain the necessary approvals, to make the order and to advertise it, the undertaker should notify the traffic authority at least one month beforehand if an order will be needed.

7.3.13 The undertaker must, at the preliminary stages of the works, submit full supporting information with his application for an order. In cases where it is known in advance that a job will last more than 18 months, the street works would clearly be a major project to which the procedures set out in Chapter 6 will apply, giving the traffic authority the additional time needed to enable them to comply with the more onerous notice and procedural requirements of the Regulations.

Continuation of closures and restrictions

7.3.14 A 5-day temporary traffic closure or restriction notice for street works purposes cannot be extended. By contrast, a 21-day temporary traffic closure or restriction notice can be extended by one further notice giving up to 21 days more. Both 5-day and 21-day temporary traffic closure or restriction notices may be followed immediately by a temporary order, which may be made without the 7 days prior notice normally needed for such orders.

7.3.15 In some cases the original estimate of the duration of the works may change, when a statutory Revised Duration Estimate notice under s74 of the Act (see Chapter 8) will need to be served, and there may be cases where they will unavoidably overrun the temporary notice period.

7.3.16 Where it is apparent from the start of the works that this will happen, undertakers must inform the traffic authority, and the authority will take the necessary follow-up action without delay so as to enable the works to continue without interruption.

7.3.17 If after works have started it becomes apparent that they will overrun the notice period, the undertaker should immediately inform the authority that either a further notice or an order would be required. This may well need to be done in advance of the serving of a statutory Revised Duration Estimate notice. In the case of a 5-day notice it may not be possible for the follow-up order to be made in time to become operative on the expiration of the notice and, where practicable, works may have to be suspended and the site temporarily restored to traffic, until the necessary procedures are complied with and the order is made. The traffic authority will, however, endeavour to reduce to a minimum the number of cases where this has to happen and, where it is unavoidable, the period of suspension involved. This problem is unlikely to arise in the case of a 21-day temporary notice.

7.3.18 Within the time limit referred to in paragraph 7.3.4 above, a closure or restriction imposed by a temporary order may be continued by a further order. Should such an order be required the undertaker should notify the traffic authority immediately, giving, wherever possible, at least one month's notice.

Policy guidance

7.3.19 When a notice or order has been made, the undertaker will be responsible for complying with the relevant requirements of the traffic authority and police for the closure of the road.

7.3.20 Undertakers are required by section 66 of the Act to carry on and complete their street works "with all such despatch as is reasonably practicable", and street authorities may require unreasonably lengthy obstructions to be mitigated or discontinued. There is therefore a presumption that closures or restrictions will remain in force only for as long as necessary for the purposes for which they are imposed. Street authorities are, for their part, under a statutory obligation to maintain a public right of passage, and they also are expected to carry out their works with due despatch.

Charges

7.3.21 Section 76 of the Act allows for traffic authorities to recover their reasonable costs in issuing temporary notices or making temporary traffic regulation orders (TRO). Upon application for a TRO, highway authorities should provide utilities with the estimated cost of the order. Invoices should be itemised as follows:

1. Cost of Order
2. Cost of Advertising in London Gazette

3. Cost of Advertising in Local Paper
4. Cost of Administration

HAUC has agreed, in order to avoid local negotiations each time this provision is used, that standard charges (i.e. the costs of the order and administration, but not advertising which is charged separately at cost) should be levied in accordance with the agreed scale. This will be found posted on to the DETR Web Site website (www.detr.gov.uk), as soon as possible after the publication of this code.

7.3.23 Highway authorities are strongly urged to follow this advice. These charges will be reviewed periodically by HAUC.

7.3.24 The charges may be exceeded where the traffic authority can show that exceptional administrative work was required.

7.4 Street works licences

Background

7.4.1 The regime for licences is governed by section 50 of and Schedule 3 to the Act.

Notice requirements

7.4.2 Prior to issuing a licence, a street authority must give at least 10 days' notice to undertakers and others likely to be affected.

7.4.3 Prior to starting work, the licensee must give the notice of the exact date of starting work to the street authority, copied to undertakers and others likely to be affected.

7.4.4 Undertakers should also be notified by the authority of any change of ownership (where known to the street authority), surrender or withdrawal of a licence.

7.4.5 A 7-day notice must be given either by the street authority or the former licensee to those affected prior to any works being done to remove apparatus installed under a licence.

7.4.6 The street authority should ensure that each street works licensee, and the licensee's agent where applicable, is made aware of the obligations imposed by the Act, in relation, for instance, to:

— safety, signing, lighting and guarding,

— qualifications of operatives and supervisors,

— delays and obstructions,

— other undertakers' apparatus that might be affected,

— reinstatement,

— records of apparatus, and

— the needs of disabled people.

The street authority should inform the licensee of the restrictions that apply in relation to streets subject to special controls and relevant section 58 notices, that the street authority may direct times of working and that they may recover all of their inspection costs. The licensee should also be made aware of the new requirements in relation to the implementation of section 74 of the Act. Reinstatement specifications and guarantee periods will be exactly the same as for any other undertaker of street works. All apparatus should be laid wherever possible in conformity with NJUG publication No. 7 (obtainable from the National Joint Utilities Group, 30 Millbank, London, SW1P 4RD).

7.4.7 Relevant consent conditions can be included in the licence or permission to cover many of the above matters.

Record keeping

7.4.8 Street authorities will maintain a record of all works and apparatus installed under street works licences granted by them. In responding to plant enquiries by undertakers, street authorities they should include plans or details of such apparatus as appropriate.

7.5 Maintenance of undertakers' apparatus

Background

7.5.1 Undertakers are under an obligation under section 81 of the Act to maintain their apparatus in the street to the reasonable satisfaction of the street authority, having regard to the safety and convenience of traffic and the structure of the street and integrity of apparatus in it. Bridge, sewer and transport authorities also have an interest so far as any land, structure or apparatus of theirs is concerned.

7.5.2 Most undertakers, under their own legislation, have statutory obligations to maintain their networks — quite apart from which they must all maintain their systems in efficient working order so as to properly discharge their safety and service obligations to their customers.

7.5.3 Thus authorities and undertakers have a shared interest in the proper maintenance of apparatus in the street.

Practical considerations

7.5.4 Although the Act gives street authorities certain default powers to inspect and carry out emergency works, neither street authorities nor undertakers expect the need to arise. However, should it happen, then (without impeding any immediate emergency action which the circumstances may require) the matter will be referred to the agreed conciliation and arbitration machinery.

7.5.5 Where surface apparatus is found to be defective or the cause of significant surface irregularity, or where an unexplained subsidence or other disturbance of the road surface occurs, the street authority must immediately notify the undertaker concerned in accordance with Appendices D or E, and may arrange a joint inspection by agreement with the undertaker.

7.5.6 If the cause of the problem is agreed to be the responsibility of the undertaker, that undertaker must take immediate action to investigate the problem and initiate any necessary remedial works.

7.5.7 Any such remedial works will be subject to the normal notification procedures contained within Appendices D or E.

7.5.8 It is important that only the responsible undertaker, or a specialist contractor working on their behalf, investigates suspected damaged or defective undertaker's apparatus. Street authorities will only carry out investigations or remedial works using appropriately trained and experienced persons in the case of an emergency, or where the undertaker is unable or unwilling to use his own operatives or such specialist contractor.

7.5.9 Notification will be made using the Works_Comments facility and selecting the appropriate Comments_Type_Code for dangerous or non-dangerous. The Works_Promoters_Reference will in fact be the highway authority's own reference constructed as shown in E3.6.1, and using the authority's own two DETR characters. Where an undertaker accepts responsibility for previously unattributable works he must issue relevant notices using his own Promoter_Works_Ref, not that generated by the highway authority.

7.5.10 Where the street authority has opened the street or exposed undertakers apparatus in an emergency or in the circumstances described in paragraph 7.5.5, the undertaker will assist the authority by either jointly inspecting the problem to determine necessary remedial works or confirming approval for the authority to proceed.

7.5.11 The authority should specify the time within which it is reasonable for the undertaker to assist before the authority commences any remedial works.

7.6 Obstructions and delays

7.6.1　Section 66 of the Act requires undertakers to complete their works with all such dispatch as is reasonably practicable. Where a street authority believes that an undertaker is causing an obstruction by occupying more of the street, or taking longer to complete the works, than is reasonable, they may issue a notice, either to reduce the obstruction or to remove it altogether. If such a notice is issued, the undertaker must comply within 24 hours or any longer period specified in the notice.

7.6.2　Where an undertaker is co-operating with the street authority in their co-ordination duties, progress of street works should be discussed and agreed in such a way that unavoidable obstructions and delays are accommodated without the need for notices under section 66. Notices should therefore only be issued in circumstances where co-operation is not forthcoming and excessive obstructions or delays occur without adequate explanation by the undertaker.

7.7 Reinstatements

Notices

7.7.1　Under section 70(3) of the Act an undertaker must inform the street authority that he had completed reinstatement before the end of the next working day after the reinstatement was completed. This notification may be done electronically using the data set described in Appendix E.

7.7.2　The notification referred to in paragraph 7.7.1 must be followed up by the registration of the reinstatement (which must include the dimensions of the reinstated areas) within 7 working days.

7.7.3　It follows that, where the dimensions of the reinstatement are known before the end of the next working day after the reinstatement was completed, the registration may be done immediately, fulfilling the requirements of both paragraphs 7.7.1 and 7.7.2 in the same notification.

Permanent reinstatement

7.7.4　Where permanent reinstatement follows immediately after completion of the street works, the original notice will, of course, still apply.

7.7.5　Under section 70 of the Act, where interim reinstatement has been carried out, the undertaker must complete permanent reinstatement as soon as is reasonably practicable and in any event within 6 months (or up to 12 months if agreed with the street authority) of completion of the interim reinstatement. When returning to carry out permanent reinstatement the normal rules as to notice and registration will apply, but any additional notice given should be cross-referenced to the original notice for recording purposes.

Defective reinstatement

7.7.6   Under section 72 of the Act, a street authority has certain powers in relation to defective reinstatements, i.e. those not meeting the requirements detailed in the Code of Practice entitled "Specification for the Reinstatement of Opening in Highways" and the Street Works (Reinstatement) Regulations 1992 S.I. 1992 No. 1689.

Procedure

7.7.7   The procedure for defective reinstatements can be found in the Inspections Code of Practice.

7.7.8   The flow chart (Figure 2) on the following page illustrates the procedure.

Figure 2 Defective Reinstatement Notice Procedure

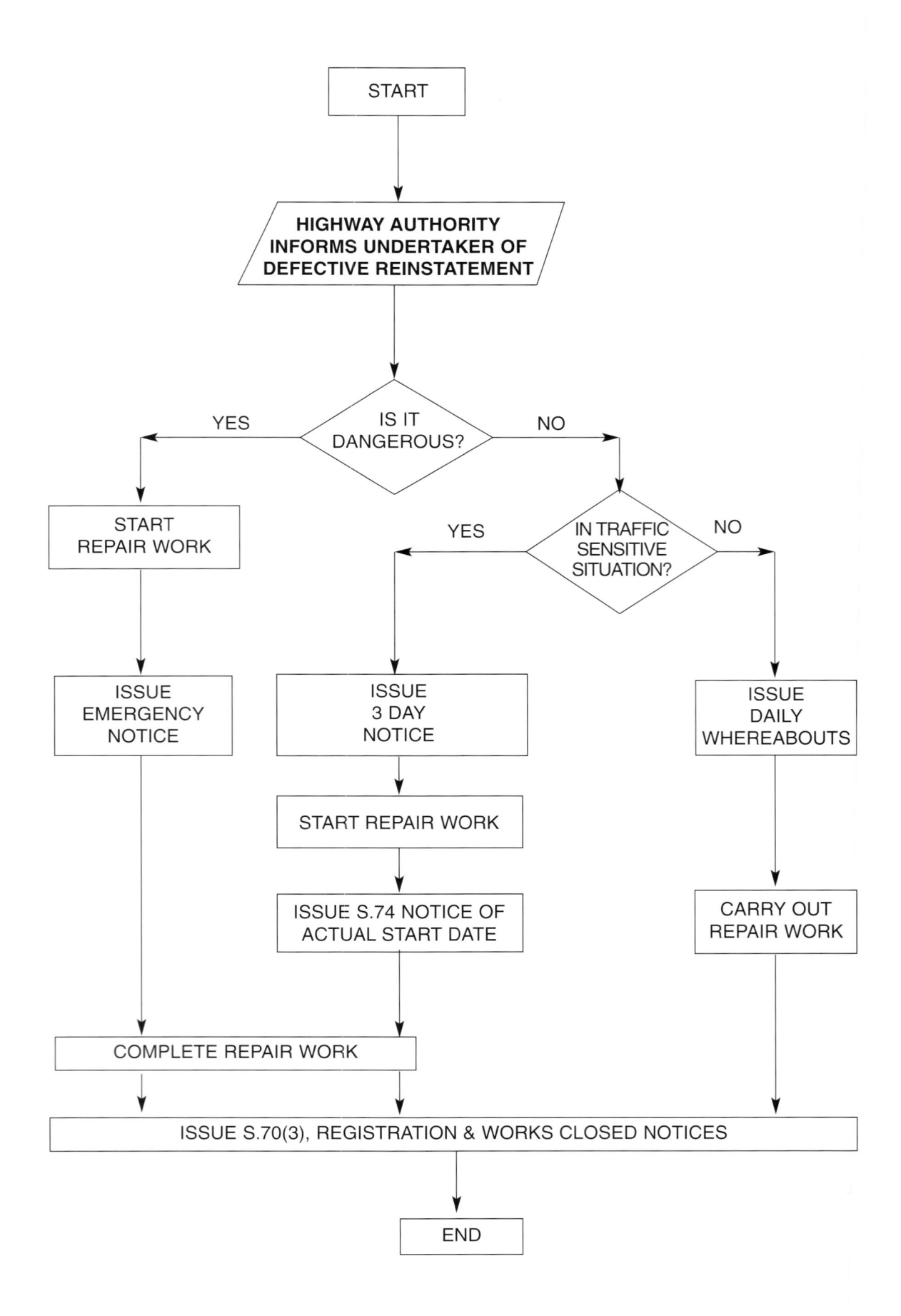

```
                        START
                          │
                          ▼
          ┌───────────────────────────────┐
          │   HIGHWAY AUTHORITY           │
          │   INFORMS UNDERTAKER OF       │
          │   DEFECTIVE REINSTATEMENT     │
          └───────────────────────────────┘
                          │
                          ▼
                    ◇ IS IT ◇
          YES ◀─── DANGEROUS? ───▶ NO
           │                        │
           ▼                        ▼
      ┌─────────┐          ◇ IN TRAFFIC ◇
      │  START  │   YES ◀── SENSITIVE ──▶ NO
      │ REPAIR  │           SITUATION?     │
      │  WORK   │              │           │
      └─────────┘              │           │
           │                   ▼           ▼
           ▼              ┌─────────┐  ┌──────────┐
      ┌─────────┐         │  ISSUE  │  │  ISSUE   │
      │  ISSUE  │         │  3 DAY  │  │  DAILY   │
      │EMERGENCY│         │ NOTICE  │  │WHEREABOUTS│
      │ NOTICE  │         └─────────┘  └──────────┘
      └─────────┘              │            │
           │                   ▼            │
           │            ┌─────────────┐     │
           │            │START REPAIR │     │
           │            │   WORK      │     │
           │            └─────────────┘     │
           │                   │            │
           │                   ▼            ▼
           │        ┌──────────────────┐ ┌──────────┐
           │        │ISSUE S.74 NOTICE │ │CARRY OUT │
           │        │OF ACTUAL START   │ │ REPAIR   │
           │        │DATE              │ │  WORK    │
           │        └──────────────────┘ └──────────┘
           ▼                   │            │
      ┌───────────────────┐    │            │
      │ COMPLETE REPAIR   │    │            │
      │     WORK          │    │            │
      └───────────────────┘    │            │
           │                   │            │
           ▼                   ▼            ▼
      ┌──────────────────────────────────────────────┐
      │ ISSUE S.70(3), REGISTRATION & WORKS CLOSED   │
      │ NOTICES                                      │
      └──────────────────────────────────────────────┘
                          │
                          ▼
                        END
```

CHAPTER 8

Section 74 — charges for unreasonably prolonged occupation of the highway

8.1 Basic Concepts

8.1.1   The regulations for the purposes of this chapter are "The Street Works (Charges for Unreasonably Prolonged Occupation of the Highway) Regulations 2001".

8.1.2   A charge may be levied upon undertakers by highway authorities where the undertaker's works in the publicly maintainable highway are unreasonably prolonged. These charges are to be retained by the highway authority to offset their costs in running the scheme. Highway authorities should inform, one month in advance of implementation the DETR and all the undertakers known to be working in their areas (see also paragraph 5.1.2) what their policy on the implementation of section 74 will be. Whether or not they are unreasonably prolonged is determined by comparing the duration of the works against the "Prescribed Period" and the "Reasonable Period". If the works take more time than the longer of these two periods, the highway authority may levy charges as detailed below. Periods are measured in whole days, part days counting as a whole day. Works not involving excavation are exempt from section 74.

8.1.3   The "Prescribed Period" is the period determined by the Secretary of State in Regulations as being the minimum period allowed for the works. Different periods may be prescribed for different classes of works. However, the Secretary of State has set the Prescribed Period as 3 days for all types of works in the initial Regulations. (Note that all references to "days" in this Chapter are references to "working days" as defined in the Act.)

8.1.4   The "Reasonable Period" is a period agreed between the undertaker and the highway authority as being reasonable for the works in question. When the undertaker gives notice of proposed works, he must state in that notice the expected duration of the works. (In practice this is achieved by giving the expected End Date, see Appendix E.) Unless the highway authority challenges this within 3 days, it becomes the Reasonable Period. At any time before the completion of the works, the undertaker may give a notice amending his estimate of the expected duration of the works. Unless the highway authority challenges this within 3 days, it becomes a new Reasonable Period, effectively overwriting the earlier one. This may happen more than once.

8.1.5   If a highway authority wishes to challenge any estimated duration given by an undertaker, it may do so by giving notice of its own estimated duration of the works. The undertaker may either accept this estimate as the Reasonable Period or, disputing it, enter into discussion with the highway authority. If no agreement can be reached by discussion on what is a Reasonable Period, then the matter will go to arbitration.

8.1.6 When the matter is resolved, the undertaker then gives another notice amending his estimate of the expected duration of the works to reflect the agreement reached which, of course, is not opposed. If the dispute has gone past the point when it is possible to issue such a notice (that is, past the expiry of the current Reasonable Period, which will be the one proposed by the highway authority – see next following sub-paragraph), then manual alterations will have to be made to records.

8.1.7 **Important Note**: until a dispute is resolved by agreement or arbitration, the works may proceed, but the highway authority's estimate of the duration of the works stands as the Reasonable Period and may be acted upon for all purposes, including raising charges. This will give rise to the possibility that a charge may have to be rebated in whole or part when the dispute is finally settled.

8.1.8 It should be noted that the Reasonable Period for Minor Works cannot exceed three days. This is because, by the definition of Minor Works, they cannot be estimated to last more than three days. Any works estimated to last longer than three days must be something other than minor works. (See Appendix B Figure B1.4 for a worked example).

8.2 Duration of Works

8.2.1 The duration of a works is measured from the date when any street works activity starts on any site in the works until the date when all street works activities are completed on all sites. Typically, the first activity will be setting out signing, lighting and guarding. It should be noted that works have not finished until all spoil, unused imported materials and any unused stores are cleared from all sites, and all signing, lighting and guarding is removed.

Interrupted Works

8.2.2 If permanent reinstatement cannot be completed on first pass, then the works will be regarded as two separate works for the purposes of section 74 charging. The first is from the start of works until the completion of interim reinstatement and the clearance of all sites. The second is from the start of permanent reinstatement until its completion and the closure of all sites. (See Appendix B Figures B1.1 and B1.2 for worked examples).

8.2.3 If remedial works are required at any site, then they will be regarded as new works for the purposes of section 74 charging, but the level of charges will relate to the definition of the original works, not to the size of the remedial works. (See Appendix B Figure B1.3 for a worked example).

8.2.4 Where an undertaker's works are interrupted because that undertaker has caused damage to a third party, then it is the responsibility of that undertaker to notify the highway authority of a revised estimated end date, taking into account the likely duration of the repair works. The site remains the responsibility of the original undertaker until he is able to issue a Works Clear or Works Closed notice.

8.3 New Notices

8.3.1 Notices given before the introduction of section 74 charging did not give the actual date of the start of a works nor the date on which a works was completed. This has meant that new notices have had to be introduced for these purposes. Other new notices are required to allow an undertaker to revise his estimate of the duration of a works and for a highway authority to challenge any estimated duration. These new notices are required under section 74 (5C) inserted into the 1991 Act by section 256 of the Transport Act 2000.

8.3.2 Fortunately, the widespread use of electronic noticing means that these new notices may be achieved with minimal changes to systems. Existing data structures will be able to handle the notices with the introduction of additional Data Capture Codes. These are given below. For giving notices on paper forms, see Appendix D.

Actual Start

8.3.3 A new notice of the Actual Start date of a works is required to start the prescribed or reasonable period. This is because the existing notices giving a proposed start date do not mean that the works have to start on the date given in the notice. Works may not begin *before* the date given in a notice of proposed works but, for example, in the case of a 7-day notice, the works may begin at any time in the 7-day period from the earliest start date given.

8.3.4 Notice of Actual Start of a works *must* be given no later than the end of the next working day after the day on which the works started.

8.3.5 Notice of Actual Start may be given using the standard Street Works Data with the following values:

NOTICE_TYPE	*Section 74*
WORKS_STATUS	*In Progress*
WORKS_START_DATE	*Actual Start Date*

Note: In the case of works only requiring notification by Daily Whereabouts, it is *not* necessary to give an Actual Start date notice. The Daily Whereabouts is to be taken as the notice of the Actual Start date. This is because works notified on a Daily Whereabouts must start on the date given. If an undertaker gives a Daily Whereabouts and for any reason does not go ahead with the works, it is essential that he cancels the notice promptly; otherwise the highway authority might assume that the works started and he could incur overrun charges. Although a Daily Whereabouts will lapse automatically, a notice of Actual Start Date will not lapse and must be cancelled. A similar logic applies to works notified on Emergency or Urgent notices. The date given in the notice should be taken as the Actual Start Date and there is no need to issue a separate Actual Start Date notice.

Revised Duration Estimate

8.3.6   In practice, notices do not give duration estimates but estimated end dates. This allows exactly the same calculations to be made.

8.3.7   An undertaker will give an estimated end date in his notice of proposed works. This will, unless challenged by the highway authority, set the Reasonable Period. If unforeseen circumstances arise, the undertaker may wish to revise his estimated end date and thus (if not challenged) revise the Reasonable Period.

8.3.8   However, a potential problem arises from the fact that works may not start on the day that the Proposed Works notice expires, see paragraph 8.3.1. If the works start on a date later than the earliest legal start date and the estimated end date is not amended, then the estimated duration is effectively shortened. This may not matter in many cases where works are planned to be of quite long duration and the undertaker will be intending to catch up the time lost by a start later than the earliest possible. However, in case of shorter duration works, the slippage in the start date may be significant. In such a case, the undertaker may revise his Estimated End Date in his Actual Start Date notice simply by giving a new Estimated End Date. Like any other estimate of the end date/duration of works, it may be challenged by the highway authority, see paragraph 8.3.12.

8.3.9   Notice of Revised Duration Estimate may be given using the standard Street Works Data with the following values:

NOTICE_TYPE	*Section 74*
WORKS_STATUS	*Proposed or In Progress*
WORKS_ESTIMATED_END_DATE	*Revised Date*

8.3.10 It is good practice for the undertaker to give a Revised Duration Estimate notice as soon as possible after the unforeseen circumstances arise which make it necessary. To do otherwise makes it more difficult for the highway authority to satisfy itself that the circumstances are genuine and may give rise to suspicion that a revised estimate is being issued simply because the works are about to overrun.

8.3.11 However, unforeseen problems may occur near the end of a works, particularly works of short duration. Therefore, a Revised Duration Estimate notice may be given at any time up until the expiry of the existing Reasonable Period, but may not be given afterwards.

Challenge to Duration Estimate

8.3.12 Within 3 days of receiving a Duration Estimate or Revised Duration Estimate a highway authority may dispute the estimate by giving notice of its own estimate to the undertaker (section 74(4) of the Act and the regulations apply).

8.3.13 Notice of Highway Authority's Duration Estimate may be given using the standard Works Comments Data with the following values:

COMMENT_TYPE_CODE	*Section 74*
RELEVANT_DATE	*The highway authority's estimate of the works end date.*
COMMENT_TEXT	*The Highway Authority must give brief reasons for the challenge*

Important Note: to enable a highway authority to assess the reasonableness of an undertaker's estimated duration, it is important that sufficient information is given in the description of the works. This must include an indication of the scale of the works and their location within the street (i.e. carriageway, footway, etc.). The description must be in plain English without any industry-specific abbreviations or jargon. For example: "Lay 250m of 150mm diameter steel pipe at 750mm of cover in the footway", gives the recipient enough to judge the reasonableness of the associated duration estimate. This requirement is part of the data validation for Works_Description_Text (see E3.6.8) and therefore if it is not followed it might give sufficient justification for a highway authority to reject a notice.

Works Clear

8.3.14 Works Clear is used following interim reinstatement. The completion of an interim reinstatement does not necessarily mark the end of that phase of a works. Works have not finished until all spoil, unused imported materials and any unused stores are cleared from all sites, and all signing, lighting and guarding is removed. This may not happen on the same day that interim reinstatement is completed.

8.3.15 Notice of Works Clear *must* be given no later than the end of the next working day following the day on which the works were clear.

8.3.16 Notice of Works Clear may be given using the standard Street Works Data with the following values:

NOTICE_TYPE	*Section 74*
WORKS_STATUS	*Works Clear*
WORKS_COMPLETE_DATE	*Date Works All Clear*

8.3.17 The introduction of another notice at this stage in the process could mean the giving of three notices following an interim reinstatement:

- the notification required under section 70(3),

- the registration of the reinstatement, and

- a Works Clear notice.

However, in many cases multiple notices may be avoided by applying the following rules:

1 If the works are clear on the same day that the interim reinstatement is completed, then the Works Clear notice must be given by the end of the next working day, and this will also serve as the section 70(3) notice.

2   If, by the time the Works Clear notice must be issued, the undertaker has the reinstatement dimensions to hand, he may give the dimensions on the Works Clear notice and this will also serve as the Registration of the Reinstatement.

3   If the works are clear on the same day that the interim reinstatement is completed, and the undertaker has the reinstatement dimensions to hand before the end of the next working day, he may give the dimensions on the Works Clear notice by the end of the next working day, and this will also serve as the section 70(3) notice and the Registration of the Reinstatement.

Works Closed

8.3.18   Works Closed is not actually a new notice type, as it has existed since the introduction of ETON. However, it has not previously been well defined and tended to be used interchangeably with Permanent Reinstatement Complete. The rest of this sub-chapter will explain its proper use.

8.3.19   Works Closed is used following permanent reinstatement. The completion of a permanent reinstatement does not necessarily mark the end of the works. Works have not finished until all spoil, unused imported materials and any unused stores are cleared from all sites, and all signing, lighting and guarding is removed. This may not happen on the same day that permanent reinstatement is completed.

8.3.20   Notice of Works Closed *must* be given no later than the end of the next working day following the day on which the works were closed.

8.3.21   Notice of Works Closed may be given using the standard Street Works Data with the following values:

NOTICE_TYPE   *Section 74*
WORKS_STATUS   *Works Closed*
WORKS_COMPLETE_DATE   *Date Works Permanently All Closed*

8.3.22   The same rules apply to permanent reinstatements as to interim reinstatements as set down in paragraph 8.3.17 above.

Informal Warning

8.3.23   Highway authorities may wish to consider sending undertakers an informal warning when their works have begun to attract overrun charges. This could be done by sending a suitably worded Comment (see E3.7). This could be automated on highway authorities' systems.

8.4 Charging where Works are Unreasonably Prolonged

8.4.1 In order to minimise administration costs, the new notices required to operate section 74 are only given at the Works level, not the Site level. This means that the Reasonable Period and whether or not works have overrun can only be determined at the Works, not the Site, level.

8.4.2 However, charges are to be applied to a street (details in paragraph 8.4.3 below). Accordingly, if a Works overruns, the appropriate charge will apply to *all* streets in that Works. It is probable that not all streets will have overrun, some may have been completed within the Reasonable Period. Nevertheless, the charge will apply to all streets in the works. This will act as an incentive to undertakers to organise their works more efficiently. Smaller works containing fewer streets will be easier to manage than sprawling works containing many, disconnected, streets.

Charges

8.4.3 The charges when the scheme is first introduced will be applied according to the following table. However, these charges may be amended by further regulation.

	Reinstatement Category		Prescribed Period	Reasonable Period
	Above 3	3 & 4		
Works without excavation	N/A	N/A	N/A	N/A
Minor Works	£500	£100	3 days	Max 3 days including prescribed period
Emergency Works Urgent Works	£500	£100	3 days	By notice details or other agreed period
Other Works*	£2000	£250	3 days	

Charges per day of overrun beyond notified / agreed durations

*Note: Any overrun on Remedial Works will be charged at the rate appropriate to the works category of the original works.

Reinstatement Category

8.4.4   Charges vary according to the type of work and the reinstatement category of the street. Reinstatement category has been used as an approximation to the importance of the street. In order to determine the reinstatement category, that given in the Additional Street Data to the National Street Gazetteer (NSG) as maintained by the NSG concessionaire, is treated as definitive.

8.4.5   Note that if works are carried out in a grass verge, footway or any other part of the highway other than the carriageway, it is the reinstatement category of the adjacent carriageway that determines the rate of charge for any overrun. Charges do not apply where the maintainable highway does not include a carriageway.

8.4.6   Should a charge occur in a street with more than one reinstatement category, the initial calculation of the charge would be at the highest category. This will stand unless challenged by the undertaker who will have to show that the whole of his works was in the area of the street in the lower reinstatement category-charging band.

8.4.7   If a highway authority has not entered its streets reinstatement categories on the NSG, they will all be treated as category 4 for the purposes of overrun charging. Highway authorities are therefore urged to ensure that they have the most up-to-date information available on the NSG.

8.5 Notices for Works on a Junction

8.5.1   The introduction of the reinstatement category as one of the determinants of the overrun charge has led to the need to clarify the rules on giving notices for works at a junction.

8.5.2   Worked example

Consider the following street plan:

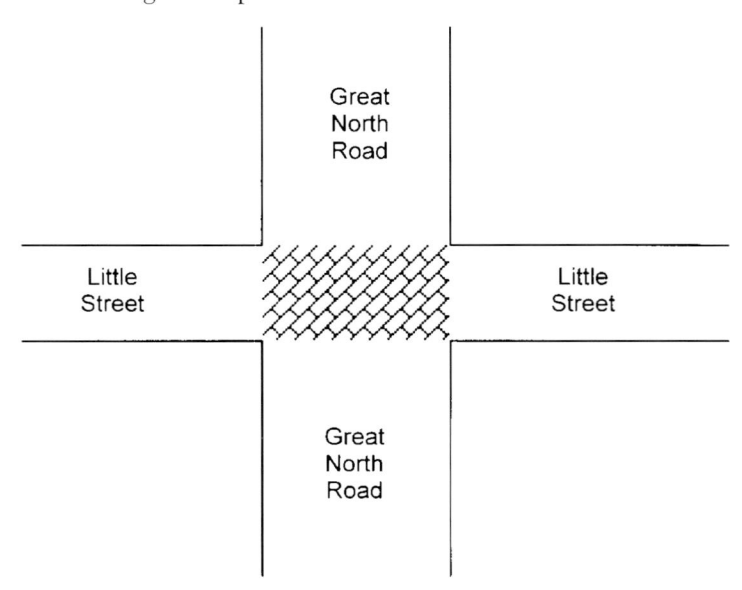

Where an undertaker proposes to do works in the central area marked in a herringbone pattern, it would appear that he could give notice of works in either Little Street or Great North Road. However, if Little Street was a category 4 (say) road and Great North Road was a category 2 (say) road then it would make a difference to any overrun charges that he might incur, which street name he chose to give notice against.

The following rule should be applied: when working in any area which could be regarded as belonging to more than one street, notice should be given against the street with the highest (0 being the highest, 4 the lowest) reinstatement category.

Note: Reinstatement Category Type 0 does not exist at the time of the publication of this Code, but it is understood that it is planned to include it as a new category in the revised *Specification for the Reinstatement of Openings in Highways*.

CHAPTER 9
Conciliation and arbitration

9.1 Introduction

9.1.1 This Code is intended to provide sufficiently detailed guidance to enable agreement on its operation and implementation to be reached at local level. Authorities and undertakers should always use their best endeavours to achieve a solution to disputes without having to refer them to conciliation. This might be achieved by referring the issue to management for settlement.

9.2 Conciliation

9.2.1 If, however, agreement cannot be reached on any matter arising under any part of this Code, except agreements of traffic-sensitivity (see paragraph 3.4.2), the dispute must be referred to conciliation, on the following basis:

(a) The issue will be reported to the regional, or, if there is one, local HAUC.

(b) Wherever possible, the conciliation should be undertaken by one person; accordingly, an independent conciliator acceptable to both parties must be appointed from a panel of local utility senior managers, local authority chief officers and nationally agreed consultants established by the regional or, if there is one, local HAUC.

(c) In matters of particular complexity, a Conciliation Panel of three persons may be set up. Each party to the dispute may nominate one member of the panel, with an independent Chairman being appointed jointly by the parties.

(d) Normally, each party must bear its own costs of conciliation, and any fees or expenses of the Conciliator or Chairman of the Conciliation Panel must be borne equally by the parties:, except where the Conciliator or Chairman considers that one party has presented a frivolous case, in which case he or she may award all costs against them.

(e) Each party must make available to the Conciliator or Panel all costing, technical and other information relevant to the matter in dispute.

(f) The conciliation must take place within two months from the date on which the issue is referred to the local (or regional) HAUC.

(g) The panel of conciliators must be agreed annually by the local (or regional) HAUC.

9.3 Arbitration

9.3.1   If an important point of principle, or a particularly expensive scheme, is involved, either party may refer the matter to arbitration, as if it were a matter to be settled by arbitration under section 99 of the Act.

APPENDIX A
Definitions

(NOTE: References in this Glossary to numbered sections are to sections of the New Roads and Street Works Act 1991, unless otherwise indicated.)

In this Code of Practice:

1	abandoned	A works may be abandoned if no street works activities have been carried out and the status of the works and the status of all the sites are "proposed". Following a works status of abandoned no further work can be carried out on the unique works reference, except in the case of Permanent Reinstatement Proposed and Remedial Reinstatement Proposed when it is permissible to reuse the original Promoter_Works_Reference. A site within a works can be abandoned if no street works activities have been carried out on the site.
2	apparatus	includes any structure for the lodging therein of apparatus or for gaining access to apparatus (section 105).
3	ASD	means "Additional Street Data" (sometimes also called "Associated Street Data"). This is data additional to the basic Gazetteer data (NSG) as defined in BS7666. The additional data is essential data relating to the requirements of the Act.
4	bridge	includes so much of any street as gives access to the bridge, and any embankment, retaining wall or other work or substance supporting or protecting that part of the street (section 88).
5	bridge authority	means the authority, body or person in whom a bridge is vested (section 88).
6	bridleway	means a highway over which the public have rights of way on foot, on horseback or leading a horse (Highways Act 1980 section 329).
7	cancel/cancelled	a notice is deemed to be cancelled if, and only if, the works to which it relates are abandoned.
8	carriageway	means that part of the highway other than a cycle track, set aside for the passage of vehicles (based upon section 329 of the Highways Act 1980).

9 culvert	means a structure in the form of a large pipe or pipes, box or enclosed channel generally used for conveying water under a road.	
10 cycle track	means a way constituting or comprised in a highway, being a way over which the public have a right of way on pedal cycles only with or without a right of way on foot (based upon section 329 of the Highways Act 1980).	
11 day	means working day, which is a day other than a Saturday, Sunday, Christmas Day, Good Friday, Bank Holiday or other prescribed public holiday. Bank Holiday means a day which is a Bank Holiday under the Banking and Financial Dealings Act 1971 in the locality in which the works in question are situated. A notice given after 4.30pm on a working day is to be treated as given on the next working day. In reckoning any period which is expressed to be a period from or before a given date, that date shall must be excluded (based upon section 98).	
12 emergency works	means works whose execution at the time when they are executed is required in order to put an end to, or to prevent the occurrence of, circumstances then existing or imminent (or which the person responsible for the works believes on reasonable grounds to be existing or imminent) which are likely to cause danger to persons or property. The term also includes works not falling within that definition but which cannot reasonably be separated from them, such as street works not at the emergency site, carried out in order to shut off or divert a supply (based upon section 52).	
13 end of the working day	means 1630 hrs (section 98).	
14 footpath	means a way over which the public have a right of way on foot only, not being a footway (section 329 of the Highways Act 1980).	
15 footway	means a way comprised in a highway which also comprises a carriageway, being a way over which the public has a right of way on foot only (section 329 of the Highways Act 1980).	
16 highway	includes the carriageway, verge and footway.	

Appendix A — Definitions

17	highway authority	means,

    (a) in the case of trunk roads (which include most motorways), the Secretary of State for Transport;

    (b) in the case of the GLA Strategic Road Network, Transport for London; and

    (c) in the case of all other roads maintainable at the public expense, the County Council, Metropolitan District Council, Unitary Authority, London Borough Council, or Common Council of the City of London in whose area the road is situated.

18 interim reinstatement complete — At the site level this means that the reinstatement has been completed to its interim state. At the works level this means that all the sites on the works have reached a stage of interim reinstatement complete. This does not mean that all activity has been finished, for example there may still be spoil and signing, lighting and guarding to be removed (see also "works clear").

19 local street gazetteer — is a subset of the National Street Gazetteer containing details of all streets in a local highway authority area. Each LSG is a self-contained entity created and maintained by the local highway authority covering all streets in their geographic area regardless of maintenance responsibility.

20 maintainable highway — means a highway which for the purposes of the Highways Act 1980 is maintainable at the public expense (section 86).

21 major projects — are projects which have been identified specifically in the annual operating programme of the undertaker or highway authority, or which, though not specifically identified in such programme, would normally be planned to commence at least 6 months in advance.

22 minor works — means works (not being emergency works or urgent works) whether in the footway, verge or carriageway, which are of a planned duration of not more than 3 days, do not form part of a rolling programme and do not involve at any one time more than 30 metres of works or leave less than the minimum width of carriageway necessary for one-way traffic (see section 5.4.1 of this Code).

23 month — means a calendar month; and a month's notice expiring on a day other than a working day is deemed to expire on the next working day.

24	National Street Gazetteer	is an application that uses BS7666 Part 1 as its standard. It is a database of street information created to the same specification. It is defined as "an index of streets and their geographical locations created and maintained by the local highway authorities". These guidelines are based on the BS7666 standard and provide additional details on the creation and maintenance of the National Street Gazetteer.
25	National Street Gazetteer concessionaire	is the organisation appointed to handle all local street gazetteers.
26	permanent reinstatement complete	At the site level this means that the reinstatement has been completed to its permanent state and the guarantee period commences for that site. At the works level this means that all the sites on the works have reached a stage of permanent reinstatement complete. This does not mean that all activity has been finished, for example there may still be spoil and signing, lighting and guarding to be removed (see also "works closed").
27	relevant authority	means, in relation to any works in a street, the street authority and also: (a) where the works include the breaking up or opening of a public sewer in the street, the sewer authority; (b) where the street is carried or crossed by a bridge vested in a transport authority, or crosses or is crossed by any other property held or used for the purposes of a transport authority, that authority; and (c) where in any other case the street is carried or crossed by a bridge, the bridge authority (section 49).
28	sites	see Appendix D4.
29	standard works	means all works which are not emergency works, urgent works or minor works.
30	start of the working day	means 0800 hrs.
31	street	means the whole or any part of any of the following, irrespective of whether it is a thoroughfare: (a) any highway, road, lane, footway, alley or passage, (b) any square or court, and

(c) any land laid out as a way whether it is for the time being formed as a way or not;

and for the avoidance of doubt includes land on the verge of a street or between two carriageways. Where a street passes over a bridge or through a tunnel, references to the street include that bridge or tunnel (section 48 etc).

32	street authority	in relation to a street means the highway authority in the case of a maintainable highway, or if the street is not a maintainable highway, the street managers (section 49).
33	street managers	means in relation to a street which is not a maintainable highway, the authority, body or person liable to the public to maintain or repair the street, or, if there is none, any authority, body or person having the management or control of the street (section 49).
34	street works	means works of any of the following kinds executed in a street, in pursuance of a statutory right or a street works licence:

(a) placing apparatus, or

(b) inspecting, maintaining, adjusting, repairing, altering or renewing apparatus, changing the position of apparatus or removing it, or works required for, or incidental to, any such works (including, in particular, breaking up or opening the street, or any sewer, drain or tunnel under it, or tunnelling or boring under the street) (section 48).

35	street works licence	means a licence granted by a street authority to a person to carry out street works.
36	traffic authority	means the highway authority for the street concerned (based upon paragraph 70 of Schedule 8 to the Act).
37	traffic-sensitive situation	means a traffic-sensitive street or that part of it which is designated traffic-sensitive, and, in the case of a limited designation, the dates or times to which the designation applies (based upon section 64).
38	undertaker	means the person in whom the statutory right to execute street works is vested, or the licensee under the relevant street works licence, as the case may be (based upon section 48).

39	urgent works	means street works (not being emergency works) whose execution at the time they are executed is required (or which the person responsible for the works believes on reasonable grounds to be required):

— to prevent or put an end to an unplanned interruption of any supply or service provided by the undertaker;

— to avoid substantial loss to the undertaker in relation to an existing service; or

— to reconnect supplies or services where the undertaker would be under a civil or criminal liability if the reconnection is delayed until after the expiration of the normal notice period;

and includes works which cannot reasonably be severed from such works (the Notices Regulations).

40	utility	means an undertaker by whom a statutory right to execute street works is exercised.
41	works	See Appendix D4.
42	works clear	the status of works clear provides a point where the promoter can declare that all information about the works has been supplied and the works are now complete. This indicates that all interim reinstatement where necessary has been done and all sites have been cleared.
43	works closed	the status of works closed provides a point where the promoter can declare that all information about the works has been supplied and the works are now complete. This indicates that all permanent reinstatement where necessary has been done and all sites have been cleared. The works can only be reopened if remedial works are required (see also "permanent reinstatement complete").
44	works for road purposes	means works of any of the following descriptions, executed in relation to a highway:

(a) works for the maintenance of the highway,

(b) any works under powers conferred by Part V of the Highways Act 1980 (improvement),

(c) the erection, maintenance, alteration or removal of traffic signs on or near the highway, or

(d) the construction of a crossing for vehicles across a footway or grass verge or the strengthening or adaptation of a footway for use as a crossing for vehicles,

or works of any corresponding description executed in relation to a street which is not a highway (section 86).

45 year   means a calendar year.

APPENDIX B
Timeline Diagrams & Flow Charts
B1 Timeline Diagrams

These diagrams are provided in order to illustrate how the noticing procedure should operate under the implementation of s74 of the Act.

Figure B1.1 **Works from Proposed to Clear**
This example assumes that the highway authority does not challenge the Reasonable Period proposed by the undertaker in his Notice of Proposed Works. The undertaker issues a Revised Duration Estimate notice. This must be done before the end of the Reasonable Period. This is also not challenged by the highway authority, thereby creating a revised Reasonable Period. Nevertheless, the Works overrun and the chargeable period is shown.

Figure B1.2 **Works from Interim Reinstatement Complete to Closed**
This example assumes that the highway authority does not challenge the Reasonable Period proposed by the undertaker in his Notice of Proposed Permanent Reinstatement. The undertaker does not issue a Revised Duration Estimate notice. The Permanent Reinstatement Works overrun the Reasonable Period and the chargeable period is shown.

Figure B1.3 **Works of Remedial Reinstatement**
This example assumes that the highway authority does not challenge the Reasonable Period proposed by the undertaker in his Notice of Proposed Remedial Reinstatement. The undertaker does not issue a Revised Duration Estimate notice. The Remedial Reinstatement Works overrun the Reasonable Period and the chargeable period is shown.

Figure B1.4 **Challenge to Duration Estimate**
This example only concerns challenges to Duration Estimates. Unnecessary detail is omitted. The example assumes that the highway authority does not challenge the Reasonable Period proposed by the undertaker in his Notice of Proposed Works. The undertaker issues a Revised Duration Estimate notice. This must be done before the end of the Reasonable Period. The highway authority challenges this by a notice in which it gives its own estimate of the Reasonable Period. There follows a discussion between the undertaker and highway authority

in which agreement is reached on a Reasonable Period between those previously proposed. The undertaker now gives another Revised Duration Estimate notice to reflect the agreement reached which, of course, is not opposed.

It should be noted that if the discussions go on too long, or the dispute has to go to Arbitration, it might be that the resolution happens after the Works are completed. In such case, matters will have proceeded on the basis of the highway authority's estimate and it will be too late to issue another Revised Duration Estimate notice. If the resolution differs from the highway authority's estimate, it may be necessary to take retrospective action including any necessary adjustment to any payment made.

Figures B1.1 to B1.3

Section 70(3) of the Act, Works Clear or Works Closed, and Registration of Reinstatement
The order in which these notices are shown on the diagrams is not the only one possible and in many cases not all three notices are needed.

If the works are clear or closed on the same day that the reinstatement is completed, then the Works Clear or Works Closed notice must be given by the end of the next working day and this will also serve as the section 70(3) notice.

If, by the time the Works Clear or Works Closed notice must be issued, the undertaker has the reinstatement dimensions to hand, he may give the dimensions on the Works Clear or Works Closed notice and this will also serve as the Registration of the Reinstatement.

If the works are clear or closed on the same day that the reinstatement is completed, and the undertaker has the reinstatement dimensions to hand before the end of the next working day, then he may give the dimensions on the Works Clear or Works Closed notice by the end of the next working day and this will also serve as the section 70(3) notice and the Registration of the Reinstatement.

Appendix B — Timeline diagrams and flow charts

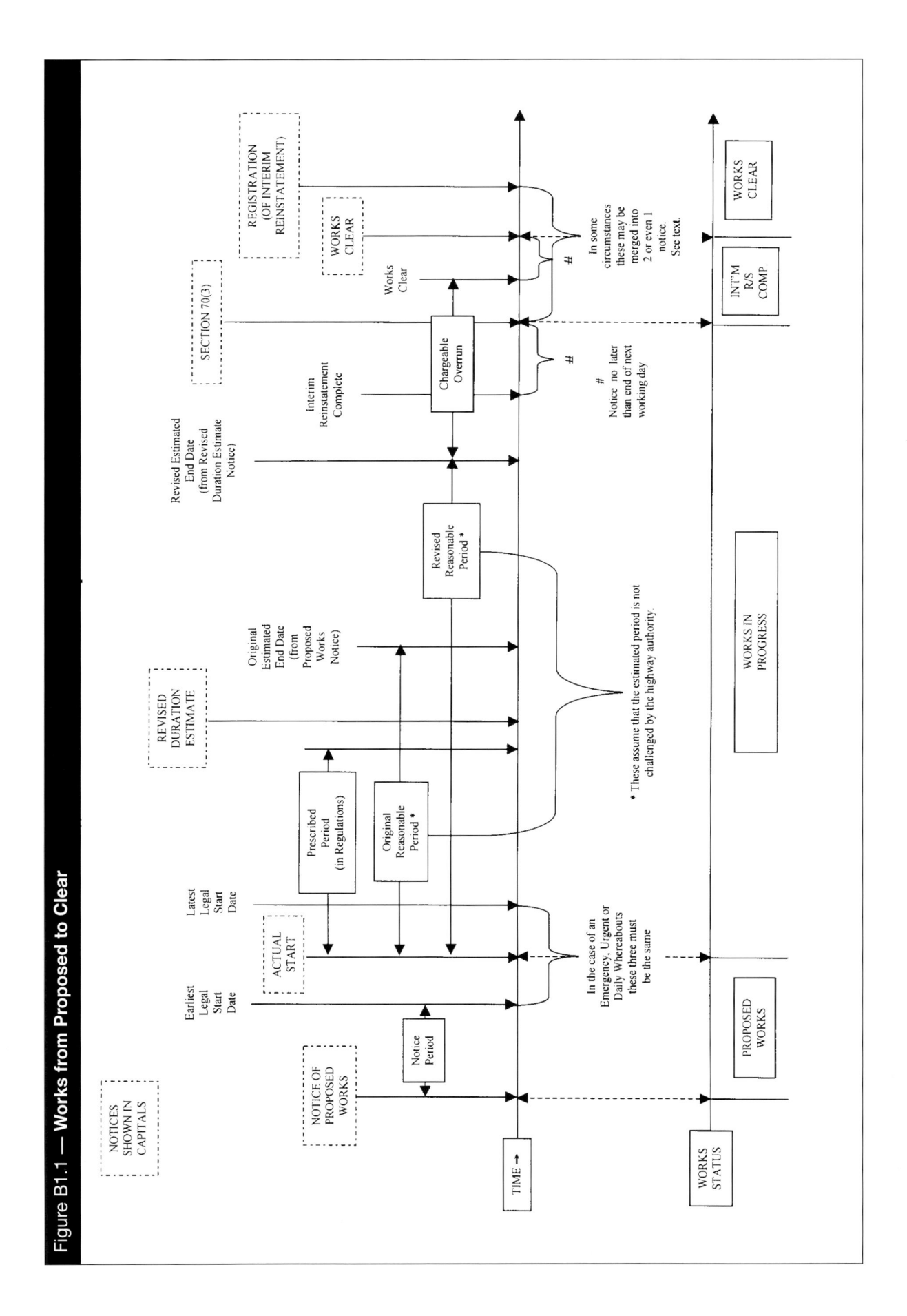

Figure B1.1 — **Works from Proposed to Clear**

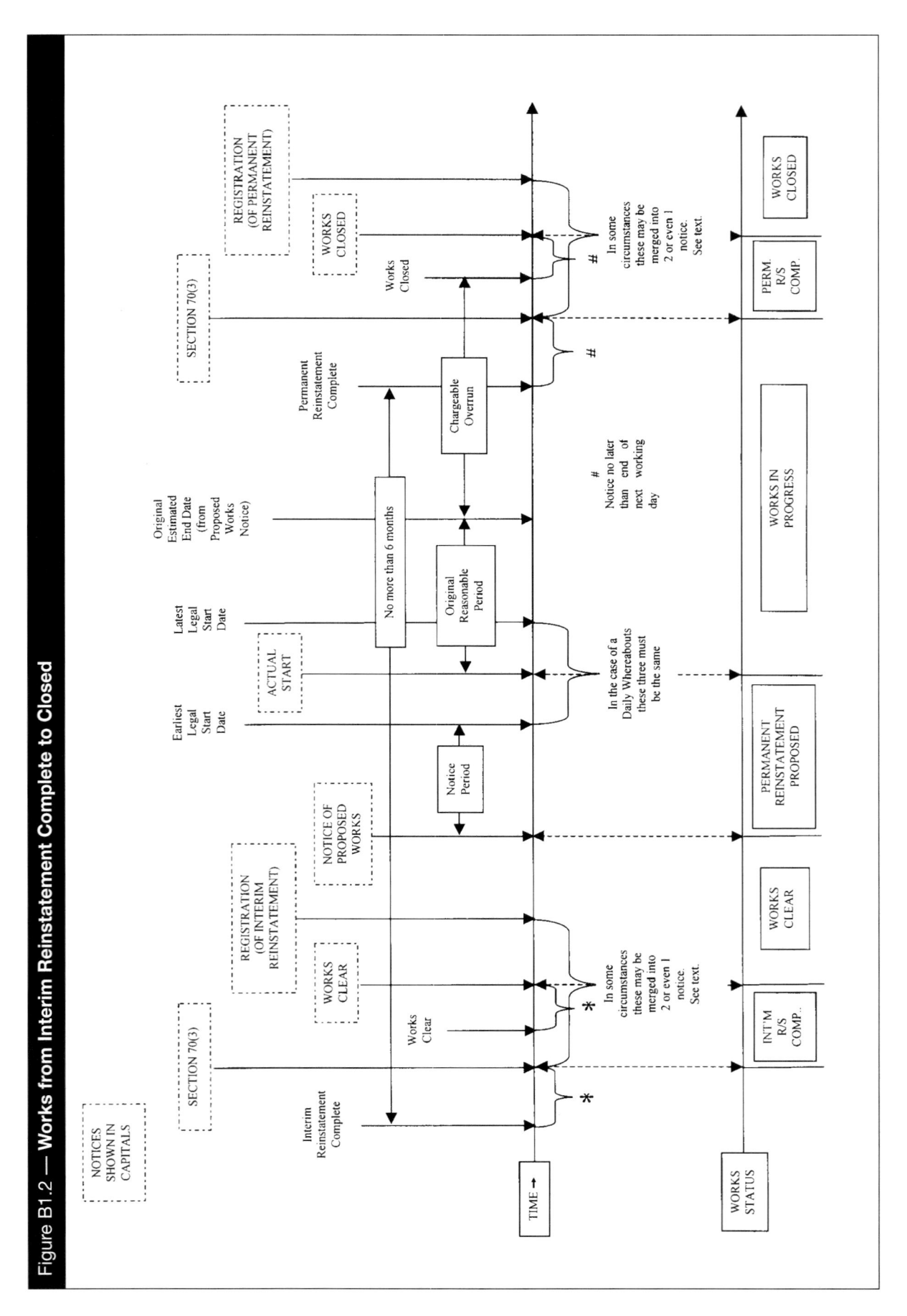

Figure B1.2 — Works from Interim Reinstatement Complete to Closed

Appendix B — Timeline diagrams and flow charts

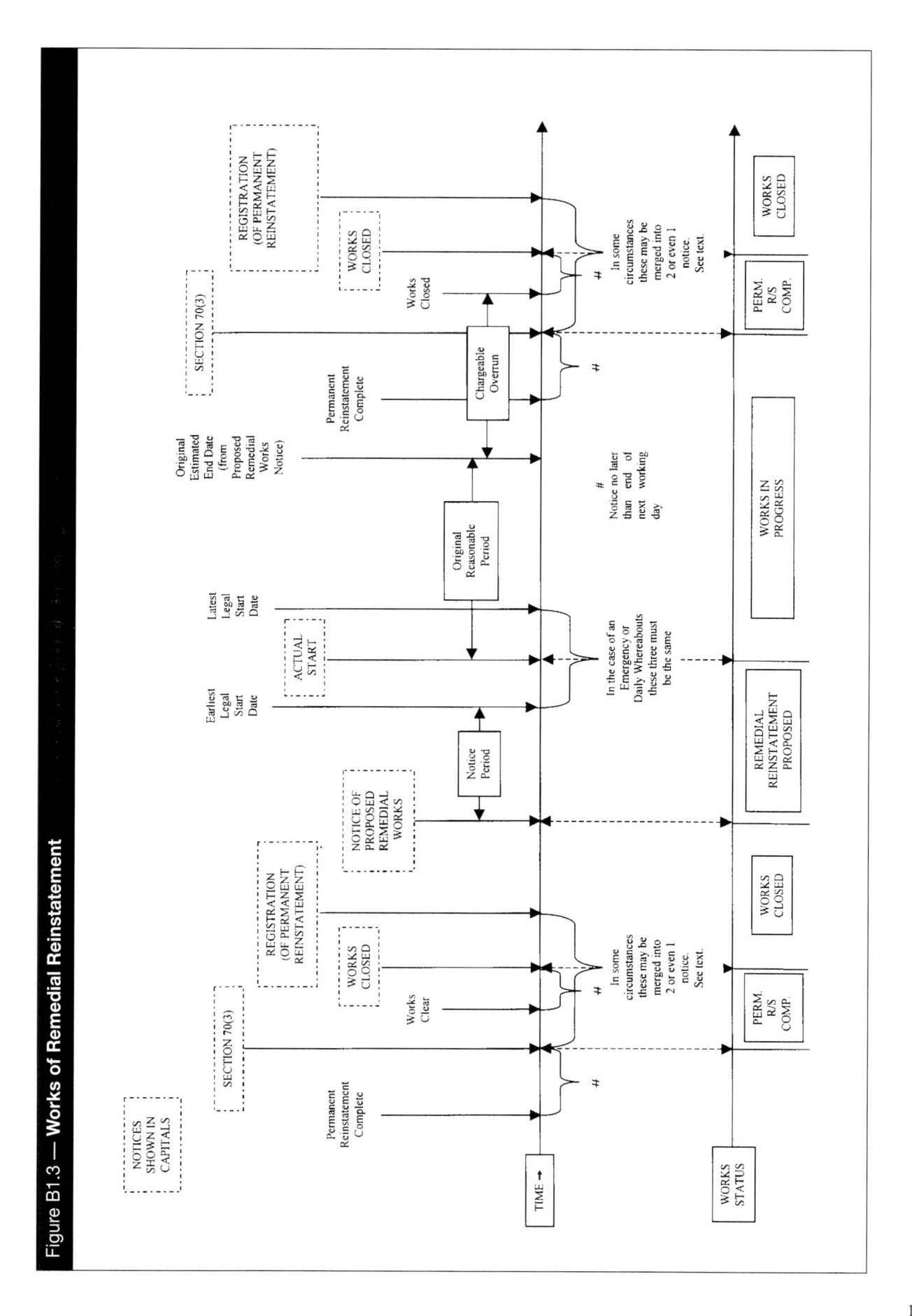

Figure B1.3 — Works of Remedial Reinstatement

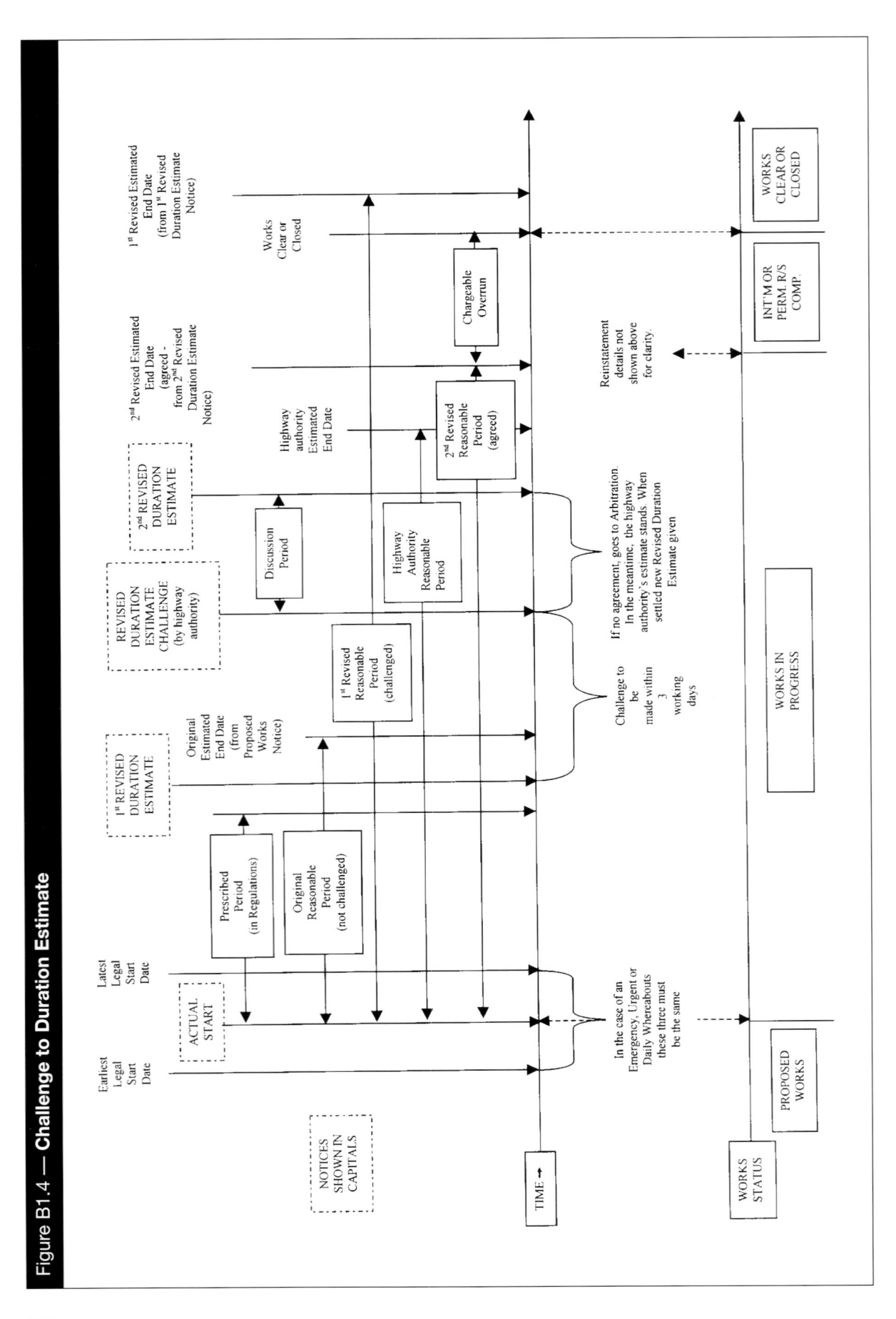

Code of Practice for the Co-ordination of Street Works etc.

Figure B1.4 — Challenge to Duration Estimate

B2 Flow Charts

The following flow charts are designed to give an indication as to the process for each category of work. They are not intended to cover all eventualities covered within these codes of practice.

Note: In order to simplify these flow charts, standard techniques have been used to show the continuation of the flow diagrams. For example, the "2" at the bottom of the page on Major Projects indicates that the process moves to the flow chart on the second page at the circle numbered "2". The arrowhead on the flow chart identifies whether the number in the circle is the start or end of a process.

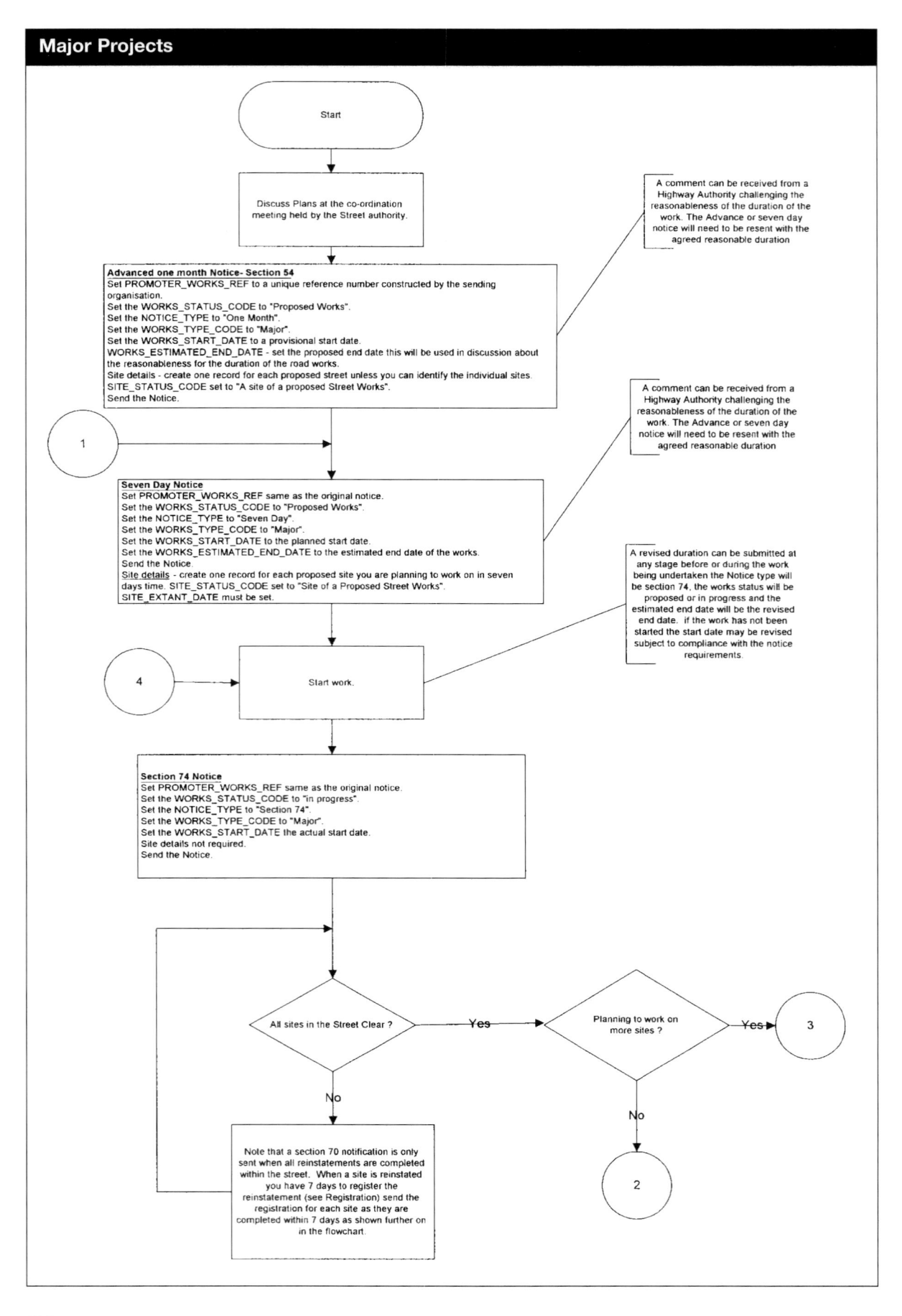

Major Projects

```
Start
  ↓
Discuss Plans at the co-ordination
meeting held by the Street authority.
  ↓
```

Advanced one month Notice- Section 54
Set PROMOTER_WORKS_REF to a unique reference number constructed by the sending organisation.
Set the WORKS_STATUS_CODE to "Proposed Works".
Set the NOTICE_TYPE to "One Month".
Set the WORKS_TYPE_CODE to "Major".
Set the WORKS_START_DATE to a provisional start date.
WORKS_ESTIMATED_END_DATE - set the proposed end date this will be used in discussion about the reasonableness for the duration of the road works.
Site details - create one record for each proposed street unless you can identify the individual sites.
SITE_STATUS_CODE set to "A site of a proposed Street Works".
Send the Notice.

A comment can be received from a Highway Authority challenging the reasonableness of the duration of the work. The Advance or seven day notice will need to be resent with the agreed reasonable duration

(1) →

Seven Day Notice
Set PROMOTER_WORKS_REF same as the original notice.
Set the WORKS_STATUS_CODE to "Proposed Works".
Set the NOTICE_TYPE to "Seven Day".
Set the WORKS_TYPE_CODE to "Major".
Set the WORKS_START_DATE to the planned start date.
Set the WORKS_ESTIMATED_END_DATE to the estimated end date of the works.
Send the Notice.
Site details - create one record for each proposed site you are planning to work on in seven days time. SITE_STATUS_CODE set to "Site of a Proposed Street Works".
SITE_EXTANT_DATE must be set.

A comment can be received from a Highway Authority challenging the reasonableness of the duration of the work. The Advance or seven day notice will need to be resent with the agreed reasonable duration

(4) → Start work.

A revised duration can be submitted at any stage before or during the work being undertaken the Notice type will be section 74, the works status will be proposed or in progress and the estimated end date will be the revised end date. if the work has not been started the start date may be revised subject to compliance with the notice requirements.

Section 74 Notice
Set PROMOTER_WORKS_REF same as the original notice.
Set the WORKS_STATUS_CODE to "in progress".
Set the NOTICE_TYPE to "Section 74".
Set the WORKS_TYPE_CODE to "Major".
Set the WORKS_START_DATE the actual start date.
Site details not required.
Send the Notice.

◇ All sites in the Street Clear ? —Yes→ ◇ Planning to work on more sites ? —Yes→ (3)

No ↓   No ↓ (2)

Note that a section 70 notification is only sent when all reinstatements are completed within the street. When a site is reinstated you have 7 days to register the reinstatement (see Registration) send the registration for each site as they are completed within 7 days as shown further on in the flowchart.

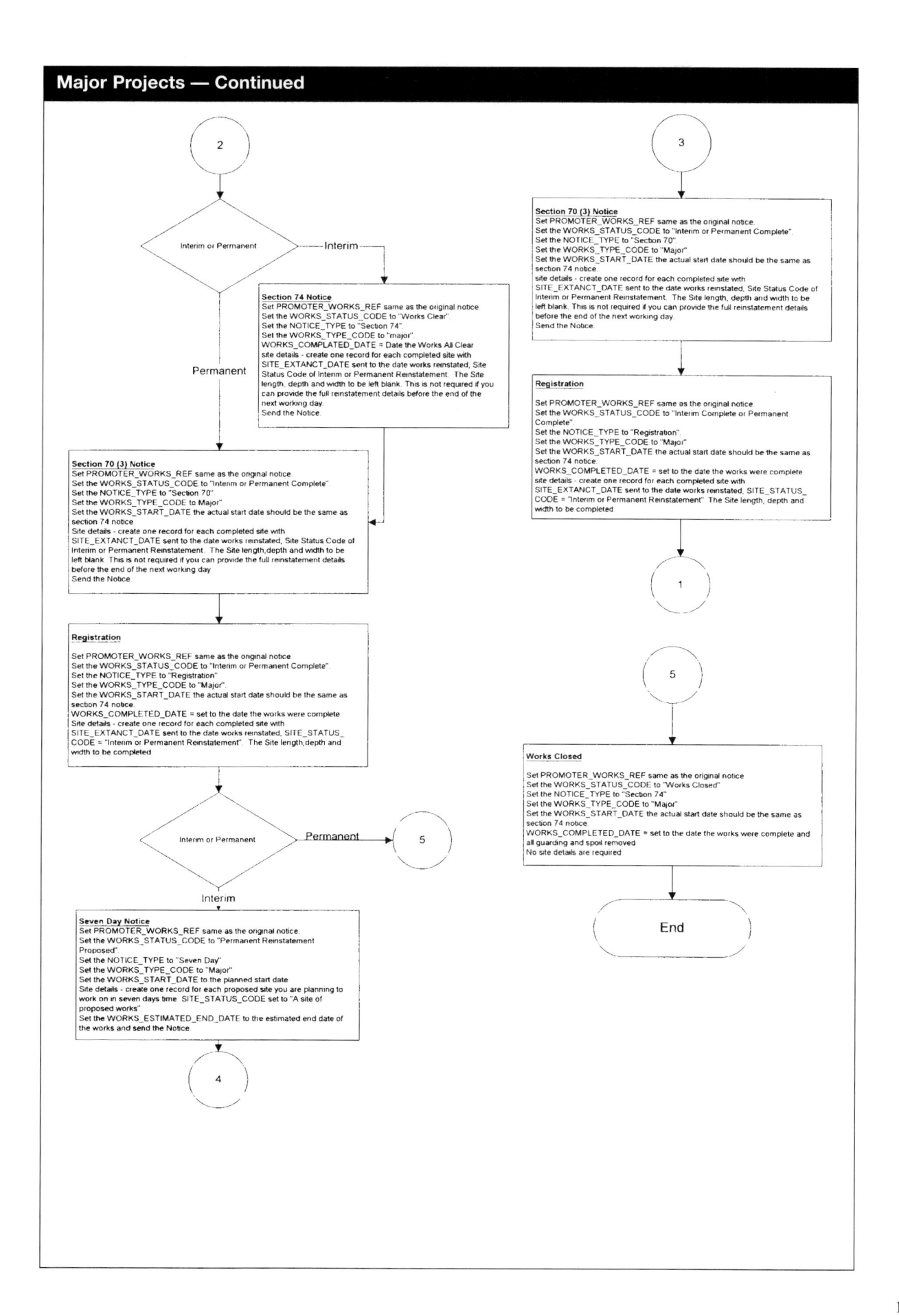

Appendix B — Timeline diagrams and flow charts

Major Projects — Continued

(2)

◇ Interim or Permanent
— Interim →

Section 74 Notice
Set PROMOTER_WORKS_REF same as the original notice.
Set the WORKS_STATUS_CODE to "Works Clear".
Set the NOTICE_TYPE to "Section 74".
Set the WORKS_TYPE_CODE to "major".
WORKS_COMPLETED_DATE = Date the Works All Clear
site details - create one record for each completed site with SITE_EXTANCT_DATE sent to the date works reinstated, Site Status Code of Interim or Permanent Reinstatement. The Site length, depth and width to be left blank. This is not required if you can provide the full reinstatement details before the end of the next working day.
Send the Notice.

— Permanent ↓

Section 70 (3) Notice
Set PROMOTER_WORKS_REF same as the original notice.
Set the WORKS_STATUS_CODE to "Interim or Permanent Complete".
Set the NOTICE_TYPE to "Section 70".
Set the WORKS_TYPE_CODE to Major.
Set the WORKS_START_DATE the actual start date should be the same as section 74 notice.
Site details - create one record for each completed site with SITE_EXTANCT_DATE sent to the date works reinstated, Site Status Code of Interim or Permanent Reinstatement. The Site length, depth and width to be left blank. This is not required if you can provide the full reinstatement details before the end of the next working day.
Send the Notice

Registration
Set PROMOTER_WORKS_REF same as the original notice.
Set the WORKS_STATUS_CODE to "Interim or Permanent Complete".
Set the NOTICE_TYPE to "Registration".
Set the WORKS_TYPE_CODE to "Major".
Set the WORKS_START_DATE the actual start date should be the same as section 74 notice.
WORKS_COMPLETED_DATE = set to the date the works were complete
Site details - create one record for each completed site with SITE_EXTANCT_DATE sent to the date works reinstated, SITE_STATUS_CODE = "Interim or Permanent Reinstatement". The Site length, depth and width to be completed.

◇ Interim or Permanent — Permanent → **(5)**
↓ Interim

Seven Day Notice
Set PROMOTER_WORKS_REF same as the original notice.
Set the WORKS_STATUS_CODE to "Permanent Reinstatement Proposed".
Set the NOTICE_TYPE to "Seven Day".
Set the WORKS_TYPE_CODE to "Major".
Set the WORKS_START_DATE to the planned start date
Site details - create one record for each proposed site you are planning to work on in seven days time. SITE_STATUS_CODE set to "A site of proposed works".
Set the WORKS_ESTIMATED_END_DATE to the estimated end date of the works and send the Notice.

(4)

(3)

Section 70 (3) Notice
Set PROMOTER_WORKS_REF same as the original notice.
Set the WORKS_STATUS_CODE to "Interim or Permanent Complete".
Set the NOTICE_TYPE to "Section 70".
Set the WORKS_TYPE_CODE to "Major".
Set the WORKS_START_DATE the actual start date should be the same as section 74 notice.
site details - create one record for each completed site with SITE_EXTANCT_DATE sent to the date works reinstated, Site Status Code of Interim or Permanent Reinstatement. The Site length, depth and width to be left blank. This is not required if you can provide the full reinstatement details before the end of the next working day.
Send the Notice.

Registration
Set PROMOTER_WORKS_REF same as the original notice
Set the WORKS_STATUS_CODE to "Interim Complete or Permanent Complete".
Set the NOTICE_TYPE to "Registration".
Set the WORKS_TYPE_CODE to "Major".
Set the WORKS_START_DATE the actual start date should be the same as section 74 notice.
WORKS_COMPLETED_DATE = set to the date the works were complete
site details - create one record for each completed site with SITE_EXTANCT_DATE sent to the date works reinstated, SITE_STATUS_CODE = "Interim or Permanent Reinstatement" The Site length, depth and width to be completed

(1)

(5)

Works Closed
Set PROMOTER_WORKS_REF same as the original notice.
Set the WORKS_STATUS_CODE to "Works Closed".
Set the NOTICE_TYPE to "Section 74".
Set the WORKS_TYPE_CODE to "Major".
Set the WORKS_START_DATE the actual start date should be the same as section 74 notice.
WORKS_COMPLETED_DATE = set to the date the works were complete and all guarding and spoil removed
No site details are required

(End)

Standard Works

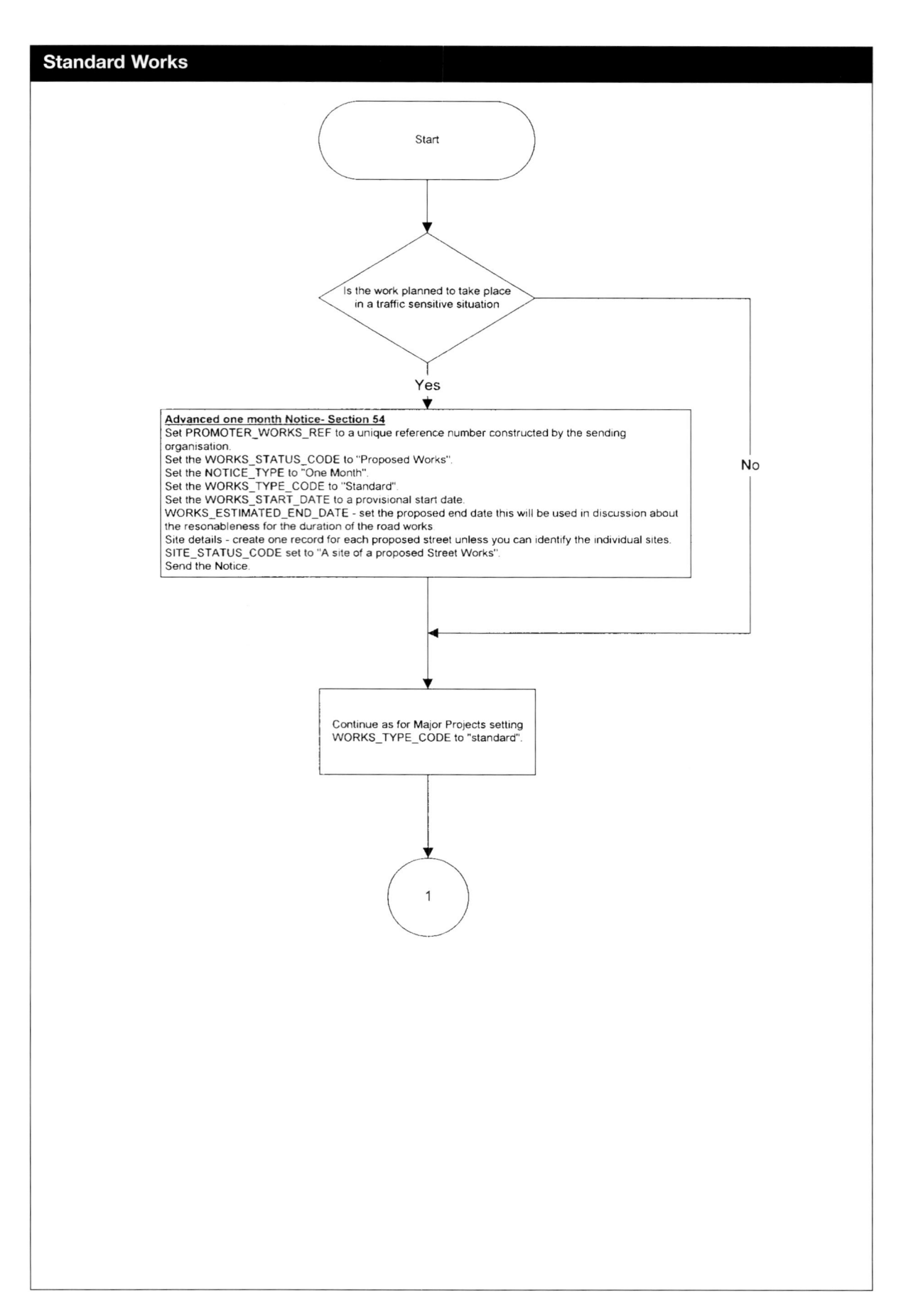

```
                    ┌─────────┐
                    │  Start  │
                    └────┬────┘
                         │
                         ▼
                    ╱ Is the work ╲
                   ╱ planned to take ╲──────────No──────────┐
                   ╲ place in a traffic ╱                    │
                    ╲ sensitive situation╱                   │
                         │                                   │
                        Yes                                  │
                         ▼                                   │
  ┌──────────────────────────────────────────────────────┐   │
  │ Advanced one month Notice- Section 54                │   │
  │ Set PROMOTER_WORKS_REF to a unique reference number  │   │
  │ constructed by the sending organisation.             │   │
  │ Set the WORKS_STATUS_CODE to "Proposed Works".       │   │
  │ Set the NOTICE_TYPE to "One Month".                  │   │
  │ Set the WORKS_TYPE_CODE to "Standard".               │   │
  │ Set the WORKS_START_DATE to a provisional start date.│   │
  │ WORKS_ESTIMATED_END_DATE - set the proposed end date │   │
  │ this will be used in discussion about the            │   │
  │ resonableness for the duration of the road works     │   │
  │ Site details - create one record for each proposed   │   │
  │ street unless you can identify the individual sites. │   │
  │ SITE_STATUS_CODE set to "A site of a proposed Street │   │
  │ Works".                                              │   │
  │ Send the Notice.                                     │   │
  └──────────────────────┬───────────────────────────────┘   │
                         │◄──────────────────────────────────┘
                         ▼
            ┌────────────────────────────┐
            │ Continue as for Major      │
            │ Projects setting           │
            │ WORKS_TYPE_CODE to         │
            │ "standard".                │
            └────────────┬───────────────┘
                         │
                         ▼
                       ( 1 )
```

Appendix B — Timeline diagrams and flow charts

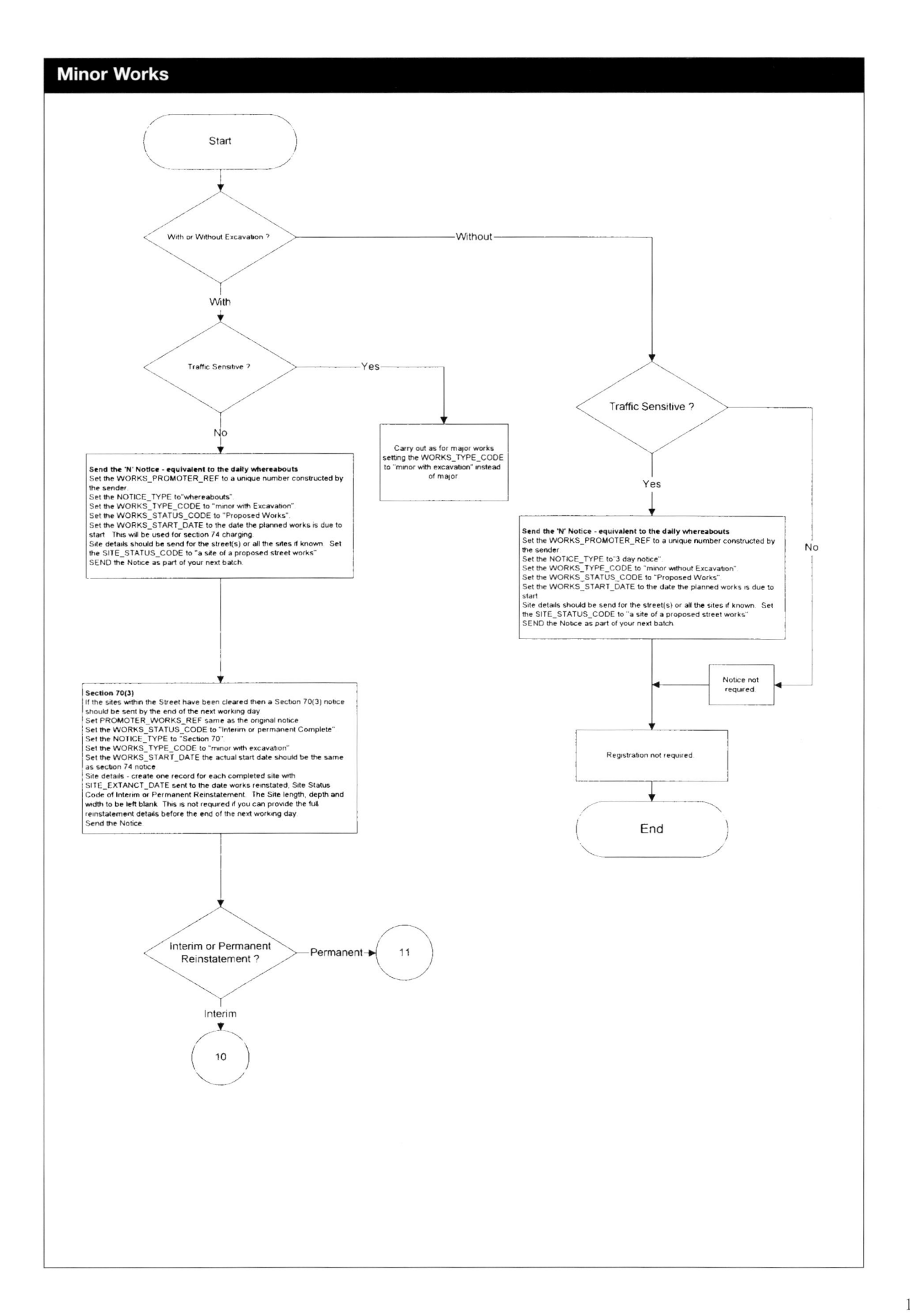

Minor Works

Start

With or Without Excavation?
- Without → (right branch)
- With ↓

Traffic Sensitive? (With branch)
- Yes → Carry out as for major works setting the WORKS_TYPE_CODE to "minor with excavation" instead of major
- No ↓

Send the 'N' Notice - equivalent to the daily whereabouts
Set the WORKS_PROMOTER_REF to a unique number constructed by the sender.
Set the NOTICE_TYPE to "whereabouts".
Set the WORKS_TYPE_CODE to "minor with Excavation".
Set the WORKS_STATUS_CODE to "Proposed Works".
Set the WORKS_START_DATE to the date the planned works is due to start. This will be used for section 74 charging.
Site details should be send for the street(s) or all the sites if known. Set the SITE_STATUS_CODE to "a site of a proposed street works".
SEND the Notice as part of your next batch.

Section 70(3)
If the sites within the Street have been cleared then a Section 70(3) notice should be sent by the end of the next working day.
Set PROMOTER_WORKS_REF same as the original notice.
Set the WORKS_STATUS_CODE to "Interim or permanent Complete".
Set the NOTICE_TYPE to "Section 70".
Set the WORKS_TYPE_CODE to "minor with excavation".
Set the WORKS_START_DATE the actual start date should be the same as section 74 notice.
Site details - create one record for each completed site with SITE_EXTANCT_DATE sent to the date works reinstated, Site Status Code of Interim or Permanent Reinstatement. The Site length, depth and width to be left blank. This is not required if you can provide the full reinstatement details before the end of the next working day.
Send the Notice.

Interim or Permanent Reinstatement?
- Permanent → (11)
- Interim → (10)

Traffic Sensitive? (Without branch)
- Yes ↓
- No → Notice not required.

Send the 'N' Notice - equivalent to the daily whereabouts
Set the WORKS_PROMOTER_REF to a unique number constructed by the sender.
Set the NOTICE_TYPE to "3 day notice".
Set the WORKS_TYPE_CODE to "minor without Excavation".
Set the WORKS_STATUS_CODE to "Proposed Works".
Set the WORKS_START_DATE to the date the planned works is due to start.
Site details should be send for the street(s) or all the sites if known. Set the SITE_STATUS_CODE to "a site of a proposed street works".
SEND the Notice as part of your next batch.

Registration not required.

End

107

Minor Works — Continued

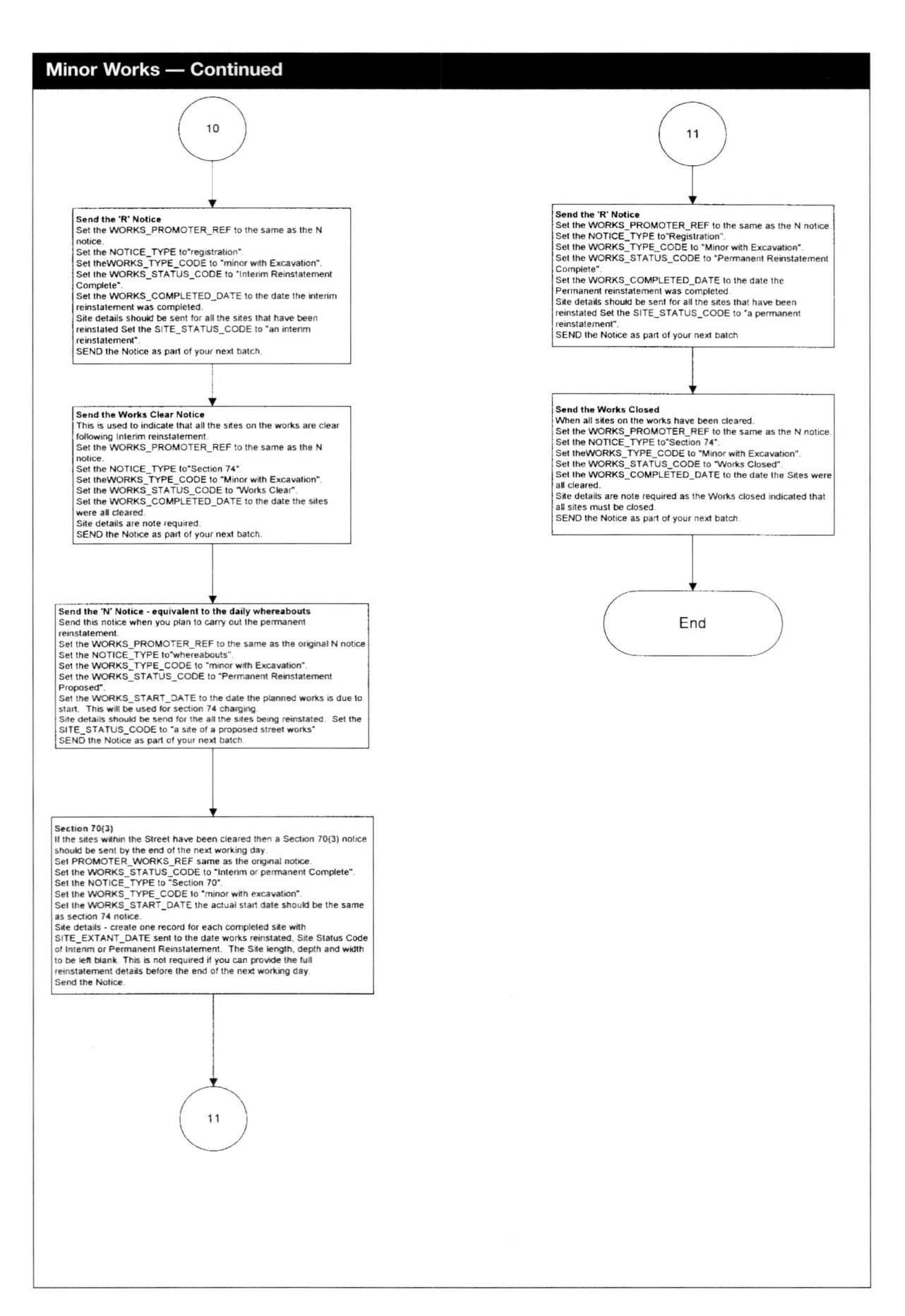

(10)

↓

Send the 'R' Notice
Set the WORKS_PROMOTER_REF to the same as the N notice.
Set the NOTICE_TYPE to "registration".
Set the WORKS_TYPE_CODE to "minor with Excavation".
Set the WORKS_STATUS_CODE to "Interim Reinstatement Complete".
Set the WORKS_COMPLETED_DATE to the date the interim reinstatement was completed.
Site details should be sent for all the sites that have been reinstated Set the SITE_STATUS_CODE to "an interim reinstatement".
SEND the Notice as part of your next batch.

↓

Send the Works Clear Notice
This is used to indicate that all the sites on the works are clear following Interim reinstatement.
Set the WORKS_PROMOTER_REF to the same as the N notice.
Set the NOTICE_TYPE to "Section 74".
Set the WORKS_TYPE_CODE to "Minor with Excavation".
Set the WORKS_STATUS_CODE to "Works Clear".
Set the WORKS_COMPLETED_DATE to the date the sites were all cleared.
Site details are note required.
SEND the Notice as part of your next batch.

↓

Send the 'N' Notice - equivalent to the daily whereabouts
Send this notice when you plan to carry out the permanent reinstatement.
Set the WORKS_PROMOTER_REF to the same as the original N notice
Set the NOTICE_TYPE to "whereabouts".
Set the WORKS_TYPE_CODE to "minor with Excavation".
Set the WORKS_STATUS_CODE to "Permanent Reinstatement Proposed".
Set the WORKS_START_DATE to the date the planned works is due to start. This will be used for section 74 charging.
Site details should be send for the all the sites being reinstated. Set the SITE_STATUS_CODE to "a site of a proposed street works"
SEND the Notice as part of your next batch.

↓

Section 70(3)
If the sites within the Street have been cleared then a Section 70(3) notice should be sent by the end of the next working day.
Set PROMOTER_WORKS_REF same as the original notice.
Set the WORKS_STATUS_CODE to "Interim or permanent Complete".
Set the NOTICE_TYPE to "Section 70".
Set the WORKS_TYPE_CODE to "minor with excavation".
Set the WORKS_START_DATE the actual start date should be the same as section 74 notice.
Site details - create one record for each completed site with SITE_EXTANT_DATE sent to the date works reinstated, Site Status Code of Interim or Permanent Reinstatement. The Site length, depth and width to be left blank. This is not required if you can provide the full reinstatement details before the end of the next working day.
Send the Notice.

↓

(11)

---

(11)

↓

Send the 'R' Notice
Set the WORKS_PROMOTER_REF to the same as the N notice.
Set the NOTICE_TYPE to "Registration".
Set the WORKS_TYPE_CODE to "Minor with Excavation".
Set the WORKS_STATUS_CODE to "Permanent Reinstatement Complete".
Set the WORKS_COMPLETED_DATE to the date the Permanent reinstatement was completed.
Site details should be sent for all the sites that have been reinstated Set the SITE_STATUS_CODE to "a permanent reinstatement".
SEND the Notice as part of your next batch.

↓

Send the Works Closed
When all sites on the works have been cleared.
Set the WORKS_PROMOTER_REF to the same as the N notice.
Set the NOTICE_TYPE to "Section 74".
Set the WORKS_TYPE_CODE to "Minor with Excavation".
Set the WORKS_STATUS_CODE to "Works Closed".
Set the WORKS_COMPLETED_DATE to the date the Sites were all cleared.
Site details are note required as the Works closed indicated that all sites must be closed.
SEND the Notice as part of your next batch.

↓

End

Appendix B — Timeline diagrams and flow charts

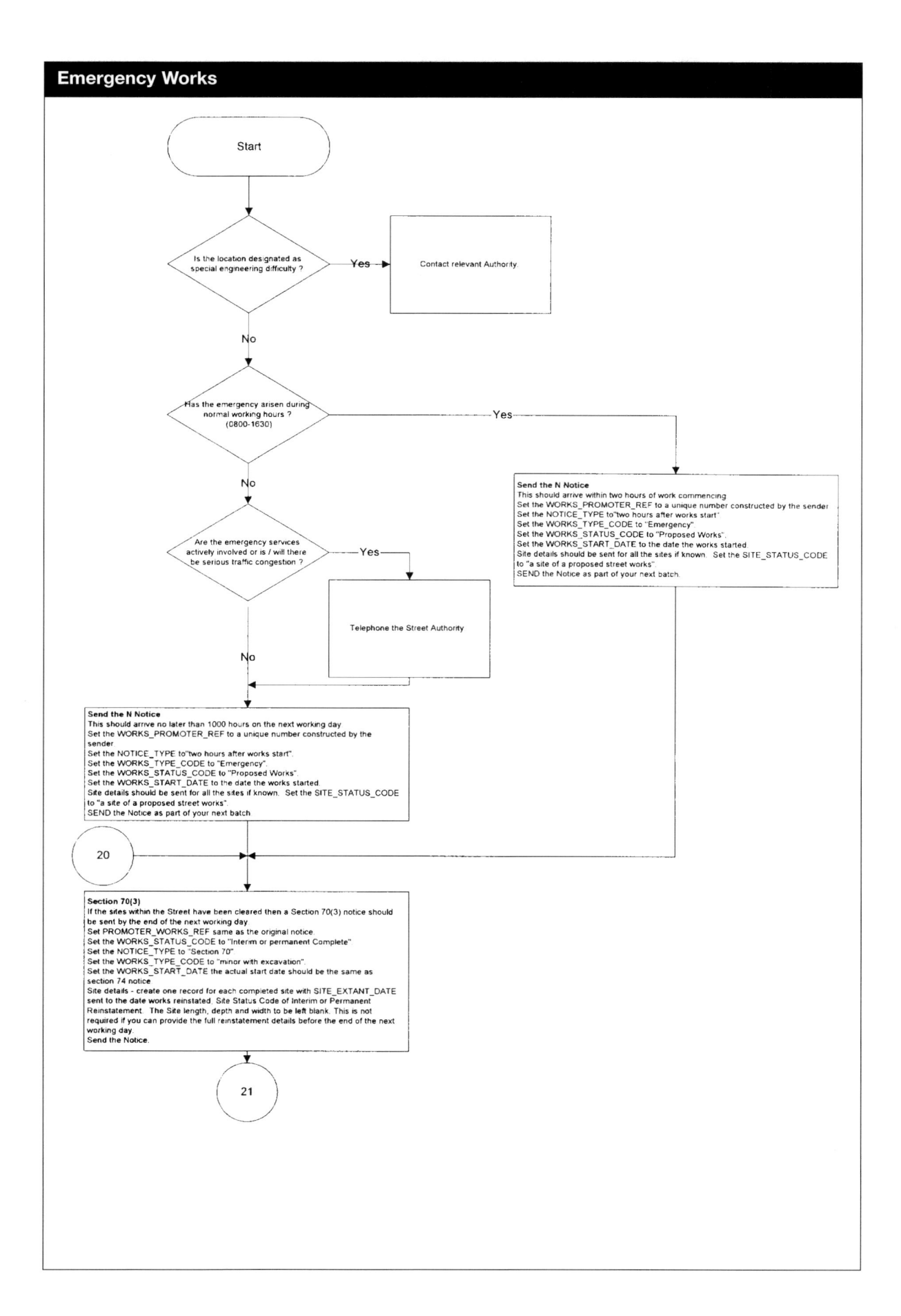

Emergency Works

```
                            Start
                              │
                              ▼
                   ◇ Is the location designated as ◇ ──Yes──▶ [Contact relevant Authority]
                     special engineering difficulty ?
                              │
                              No
                              ▼
                   ◇ Has the emergency arisen during ◇ ──Yes──────────────────────┐
                       normal working hours ?                                      │
                           (0800-1630)                                             │
                              │                                                    ▼
                              No                                    ┌──────────────────────────────────┐
                              ▼                                     │ Send the N Notice                │
                   ◇ Are the emergency services ◇ ──Yes──┐          │ This should arrive within two hours of work commencing
                     actively involved or is / will there            │ Set the WORKS_PROMOTER_REF to a unique number constructed by the sender
                     be serious traffic congestion ?                 │ Set the NOTICE_TYPE to "two hours after works start"
                              │                         ▼            │ Set the WORKS_TYPE_CODE to "Emergency"
                              No           [Telephone the Street     │ Set the WORKS_STATUS_CODE to "Proposed Works"
                              │                Authority]            │ Set the WORKS_START_DATE to the date the works started.
                              │                         │            │ Site details should be sent for all the sites if known. Set the SITE_STATUS_CODE
                              │◀────────────────────────┘            │ to "a site of a proposed street works".
                              ▼                                      │ SEND the Notice as part of your next batch.
```

Send the N Notice
This should arrive no later than 1000 hours on the next working day
Set the WORKS_PROMOTER_REF to a unique number constructed by the sender
Set the NOTICE_TYPE to "two hours after works start".
Set the WORKS_TYPE_CODE to "Emergency".
Set the WORKS_STATUS_CODE to "Proposed Works".
Set the WORKS_START_DATE to the date the works started.
Site details should be sent for all the sites if known. Set the SITE_STATUS_CODE to "a site of a proposed street works".
SEND the Notice as part of your next batch.

(20)

Section 70(3)
If the sites within the Street have been cleared then a Section 70(3) notice should be sent by the end of the next working day.
Set PROMOTER_WORKS_REF same as the original notice.
Set the WORKS_STATUS_CODE to "Interim or permanent Complete".
Set the NOTICE_TYPE to "Section 70".
Set the WORKS_TYPE_CODE to "minor with excavation".
Set the WORKS_START_DATE the actual start date should be the same as section 74 notice.
Site details - create one record for each completed site with SITE_EXTANT_DATE sent to the date works reinstated. Site Status Code of Interim or Permanent Reinstatement. The Site length, depth and width to be left blank. This is not required if you can provide the full reinstatement details before the end of the next working day.
Send the Notice.

(21)

109

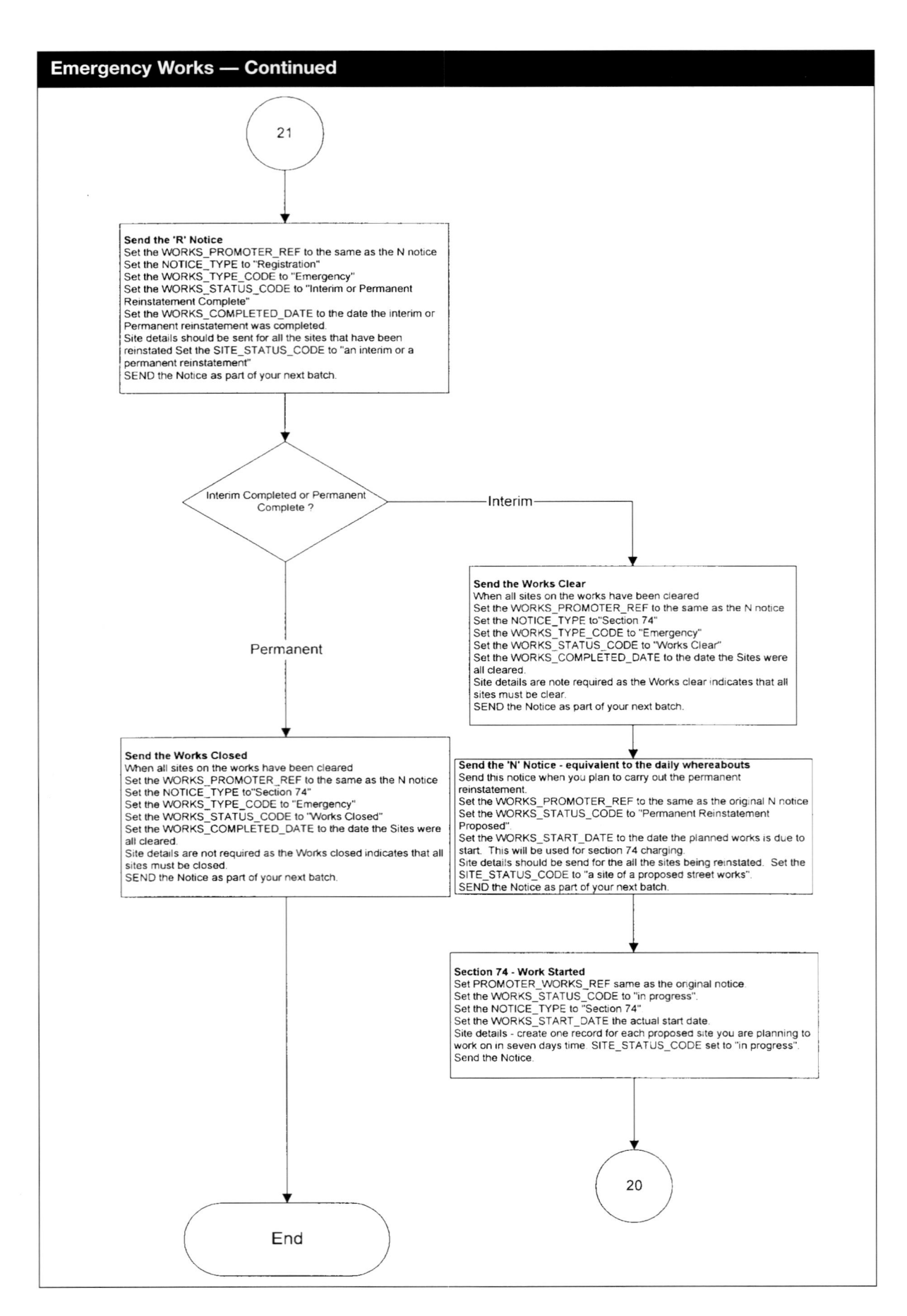

Emergency Works — Continued

(21)

Send the 'R' Notice
Set the WORKS_PROMOTER_REF to the same as the N notice
Set the NOTICE_TYPE to "Registration"
Set the WORKS_TYPE_CODE to "Emergency"
Set the WORKS_STATUS_CODE to "Interim or Permanent Reinstatement Complete"
Set the WORKS_COMPLETED_DATE to the date the interim or Permanent reinstatement was completed.
Site details should be sent for all the sites that have been reinstated Set the SITE_STATUS_CODE to "an interim or a permanent reinstatement"
SEND the Notice as part of your next batch.

Interim Completed or Permanent Complete ? — Interim →

Permanent ↓

Send the Works Clear
When all sites on the works have been cleared
Set the WORKS_PROMOTER_REF to the same as the N notice
Set the NOTICE_TYPE to "Section 74"
Set the WORKS_TYPE_CODE to "Emergency"
Set the WORKS_STATUS_CODE to "Works Clear"
Set the WORKS_COMPLETED_DATE to the date the Sites were all cleared.
Site details are note required as the Works clear indicates that all sites must be clear.
SEND the Notice as part of your next batch.

Send the Works Closed
When all sites on the works have been cleared
Set the WORKS_PROMOTER_REF to the same as the N notice
Set the NOTICE_TYPE to "Section 74"
Set the WORKS_TYPE_CODE to "Emergency"
Set the WORKS_STATUS_CODE to "Works Closed"
Set the WORKS_COMPLETED_DATE to the date the Sites were all cleared.
Site details are not required as the Works closed indicates that all sites must be closed.
SEND the Notice as part of your next batch.

Send the 'N' Notice - equivalent to the daily whereabouts
Send this notice when you plan to carry out the permanent reinstatement.
Set the WORKS_PROMOTER_REF to the same as the original N notice
Set the WORKS_STATUS_CODE to "Permanent Reinstatement Proposed".
Set the WORKS_START_DATE to the date the planned works is due to start. This will be used for section 74 charging.
Site details should be send for the all the sites being reinstated. Set the SITE_STATUS_CODE to "a site of a proposed street works".
SEND the Notice as part of your next batch.

Section 74 - Work Started
Set PROMOTER_WORKS_REF same as the original notice.
Set the WORKS_STATUS_CODE to "in progress".
Set the NOTICE_TYPE to "Section 74"
Set the WORKS_START_DATE the actual start date.
Site details - create one record for each proposed site you are planning to work on in seven days time. SITE_STATUS_CODE set to "in progress".
Send the Notice.

(20)

End

Appendix B — Timeline diagrams and flow charts

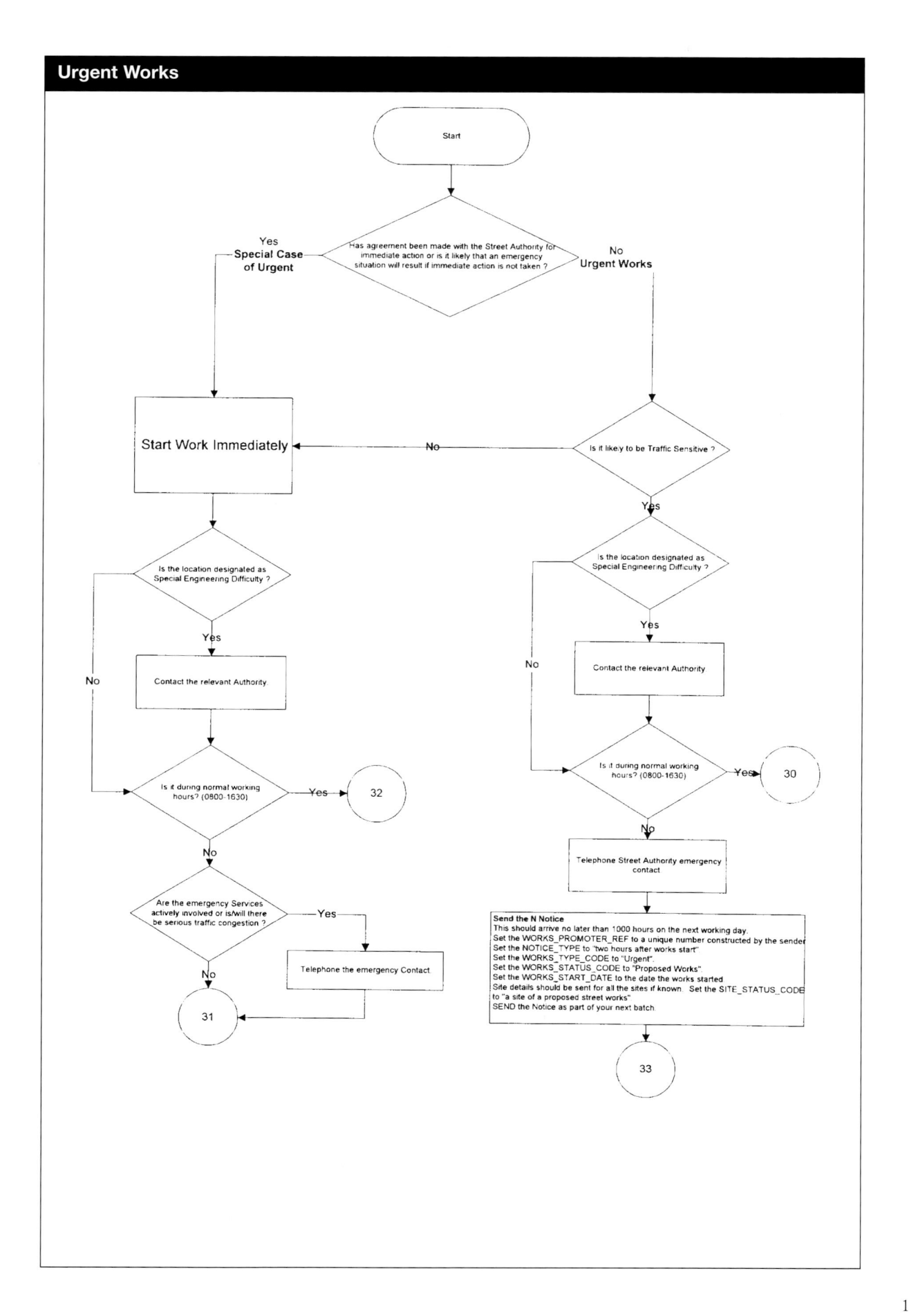

Urgent Works

Start

Has agreement been made with the Street Authority for immediate action or is it likely that an emergency situation will result if immediate action is not taken?

- **Yes** — Special Case of Urgent → Start Work Immediately
- **No** — Urgent Works → Is it likely to be Traffic Sensitive?
  - **No** → Start Work Immediately
  - **Yes** → Is the location designated as Special Engineering Difficulty?
    - **Yes** → Contact the relevant Authority
    - **No** → Is it during normal working hours? (0800-1630)
      - **Yes** → 30
      - **No** → Telephone Street Authority emergency contact

Send the N Notice
This should arrive no later than 1000 hours on the next working day.
Set the WORKS_PROMOTER_REF to a unique number constructed by the sender
Set the NOTICE_TYPE to "two hours after works start".
Set the WORKS_TYPE_CODE to "Urgent".
Set the WORKS_STATUS_CODE to "Proposed Works".
Set the WORKS_START_DATE to the date the works started.
Site details should be sent for all the sites if known. Set the SITE_STATUS_CODE to "a site of a proposed street works"
SEND the Notice as part of your next batch.

→ 33

From Start Work Immediately:
- Is the location designated as Special Engineering Difficulty?
  - **Yes** → Contact the relevant Authority
  - **No** → Is it during normal working hours? (0800-1630)
    - **Yes** → 32
    - **No** → Are the emergency Services actively involved or is/will there be serious traffic congestion?
      - **Yes** → Telephone the emergency Contact → 31
      - **No** → 31

111

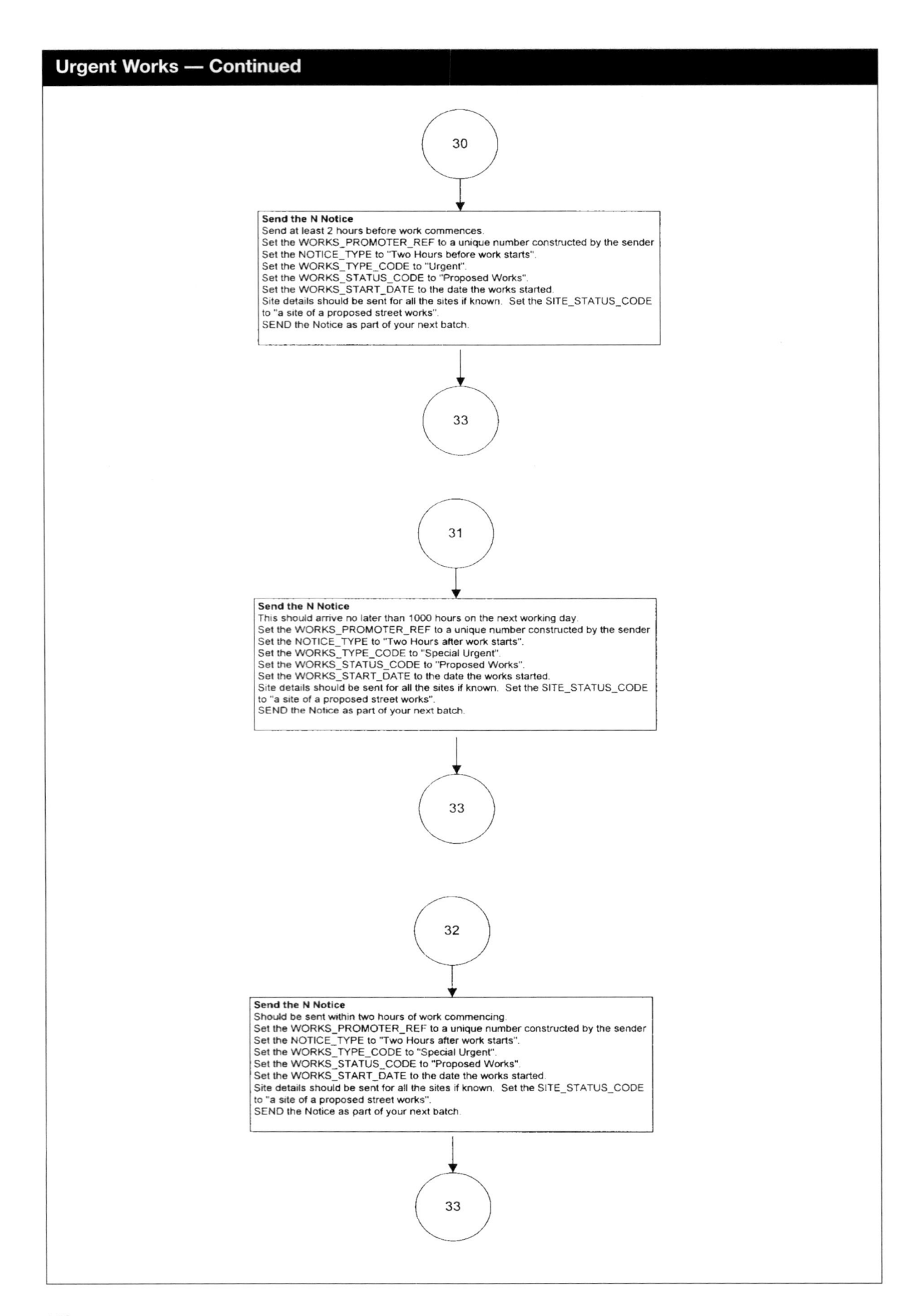

Urgent Works — Continued

(30)

Send the N Notice
Send at least 2 hours before work commences.
Set the WORKS_PROMOTER_REF to a unique number constructed by the sender
Set the NOTICE_TYPE to "Two Hours before work starts".
Set the WORKS_TYPE_CODE to "Urgent".
Set the WORKS_STATUS_CODE to "Proposed Works".
Set the WORKS_START_DATE to the date the works started.
Site details should be sent for all the sites if known. Set the SITE_STATUS_CODE to "a site of a proposed street works".
SEND the Notice as part of your next batch.

(33)

(31)

Send the N Notice
This should arrive no later than 1000 hours on the next working day.
Set the WORKS_PROMOTER_REF to a unique number constructed by the sender
Set the NOTICE_TYPE to "Two Hours after work starts".
Set the WORKS_TYPE_CODE to "Special Urgent".
Set the WORKS_STATUS_CODE to "Proposed Works".
Set the WORKS_START_DATE to the date the works started.
Site details should be sent for all the sites if known. Set the SITE_STATUS_CODE to "a site of a proposed street works".
SEND the Notice as part of your next batch.

(33)

(32)

Send the N Notice
Should be sent within two hours of work commencing.
Set the WORKS_PROMOTER_REF to a unique number constructed by the sender
Set the NOTICE_TYPE to "Two Hours after work starts".
Set the WORKS_TYPE_CODE to "Special Urgent".
Set the WORKS_STATUS_CODE to "Proposed Works".
Set the WORKS_START_DATE to the date the works started.
Site details should be sent for all the sites if known. Set the SITE_STATUS_CODE to "a site of a proposed street works".
SEND the Notice as part of your next batch.

(33)

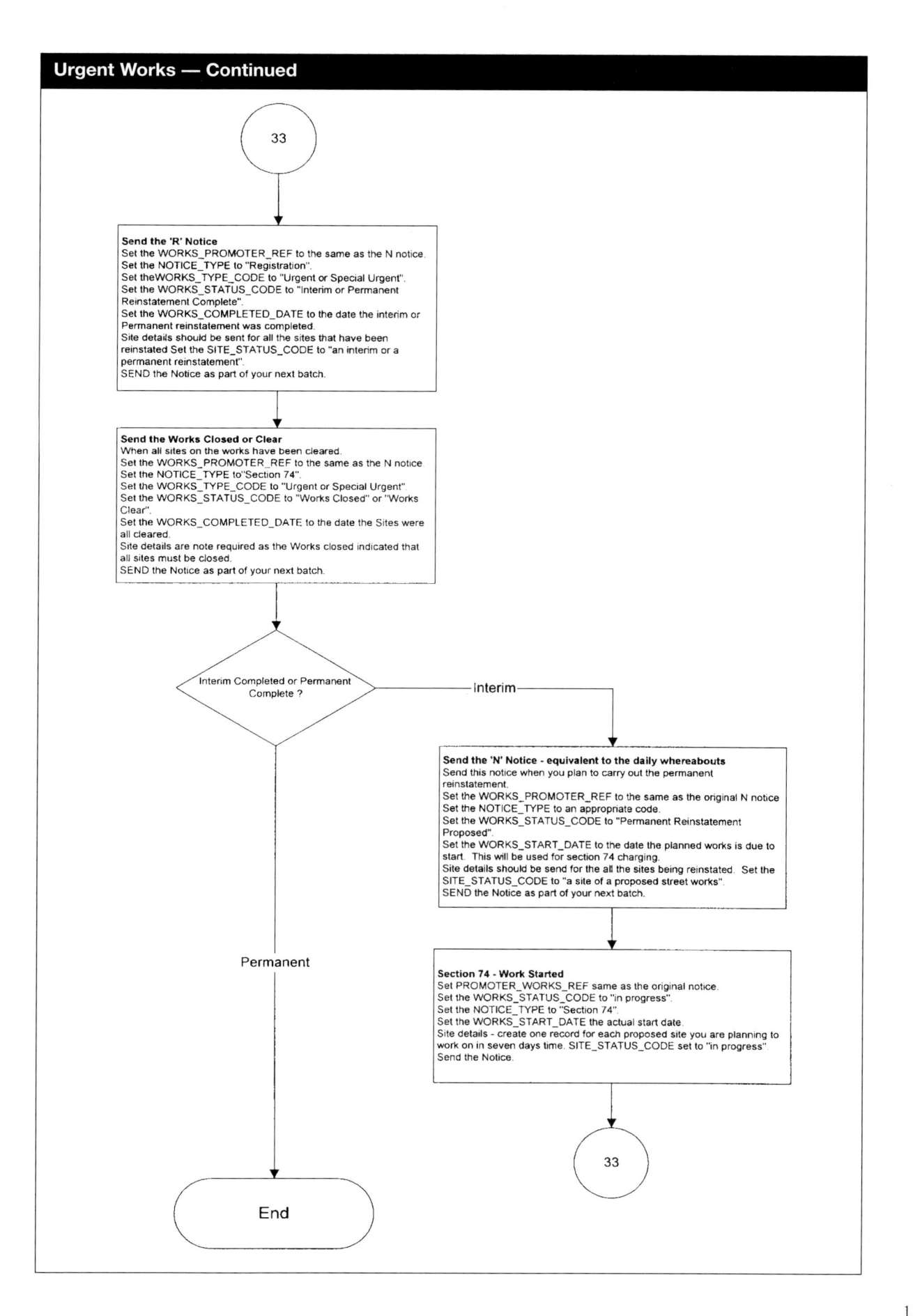

Urgent Works — Continued

(33)

Send the 'R' Notice
Set the WORKS_PROMOTER_REF to the same as the N notice.
Set the NOTICE_TYPE to "Registration".
Set theWORKS_TYPE_CODE to "Urgent or Special Urgent".
Set the WORKS_STATUS_CODE to "Interim or Permanent Reinstatement Complete".
Set the WORKS_COMPLETED_DATE to the date the interim or Permanent reinstatement was completed.
Site details should be sent for all the sites that have been reinstated Set the SITE_STATUS_CODE to "an interim or a permanent reinstatement".
SEND the Notice as part of your next batch.

Send the Works Closed or Clear
When all sites on the works have been cleared.
Set the WORKS_PROMOTER_REF to the same as the N notice
Set the NOTICE_TYPE to "Section 74".
Set the WORKS_TYPE_CODE to "Urgent or Special Urgent"
Set the WORKS_STATUS_CODE to "Works Closed" or "Works Clear".
Set the WORKS_COMPLETED_DATE to the date the Sites were all cleared.
Site details are note required as the Works closed indicated that all sites must be closed.
SEND the Notice as part of your next batch.

Interim Completed or Permanent Complete ?

— Interim →

Permanent

Send the 'N' Notice - equivalent to the daily whereabouts
Send this notice when you plan to carry out the permanent reinstatement.
Set the WORKS_PROMOTER_REF to the same as the original N notice
Set the NOTICE_TYPE to an appropriate code.
Set the WORKS_STATUS_CODE to "Permanent Reinstatement Proposed".
Set the WORKS_START_DATE to the date the planned works is due to start. This will be used for section 74 charging.
Site details should be send for the all the sites being reinstated. Set the SITE_STATUS_CODE to "a site of a proposed street works".
SEND the Notice as part of your next batch.

Section 74 - Work Started
Set PROMOTER_WORKS_REF same as the original notice.
Set the WORKS_STATUS_CODE to "in progress".
Set the NOTICE_TYPE to "Section 74".
Set the WORKS_START_DATE the actual start date.
Site details - create one record for each proposed site you are planning to work on in seven days time. SITE_STATUS_CODE set to "in progress".
Send the Notice.

(33)

End

APPENDIX C
Sample Form for the publication of Section 58 Notice of Substantial Road Works

NEW ROADS AND STREET WORKS ACT 1991
SECTION 58
NOTICE OF SUBSTANTIAL ROAD WORKS FOR ROAD PURPOSES.

1. Three months advance notice is hereby given that .... (Highway Authority) intends to carry out substantial works for road purposes.

2. The street in which these works will take place is: (describe street and indicate carriageway/footway/footpath/verge)

3. Work is expected to affect the carriageway/footway/footpath/verge between:

4. Work is expected to commence on: ..../......../...... (work should begin within one month of this date)

5. The works notified will include the following:

   (a) (detail area of work)

   (b) (detail extent of work)

6. Subject to the exemptions in the Act and Regulations under it, no street works may be executed for a period of 12 months from completion of the works described in paragraph 5 without the consent of the said highway authority, which is not to be unreasonably withheld.

Contact point:

(name)

(title)

Date:......../........./.....

APPENDIX D
Paper Transfer of Notices

D1 Introduction

D1.1 Regulations provide a variety of methods of service of notices. In cases of emergency and urgent works, notices may be delivered or sent by fax or electronic means, or by other agreed means. In other cases, first class post is an additional option. Since the introduction of Electronic Transfer of Notices (ETON, see Appendix E) the vast majority of notices have been given electronically. However, users are required to retain the facility of receiving notices by the other means mentioned above.

D1.2 In order to provide for a standardisation of notice forms, which will incorporate details given by Daily Whereabouts, 'N' and 'R', the process has been streamlined to utilise a paper notice similar to an Appendix E produced notice. This will enable the general rules for serving notices to be more easily applied to paper and electronic service as problems with the original format of paper notices included:

— failure of the data structure to match ETON;

— the lack of nationally unique reference numbers; and

— the lack of USRNs.

It was therefore decided to abandon the "N" and "R" forms and introduce forms compatible with ETON (see below for details).

D1.3 As well as overcoming the problems mentioned above, these new forms accommodate the new notices required by the introduction of section 74 charging without introducing a multiplicity of forms.

D2 Prerequisites for giving a notice using paper

D2.1 The undertaker will have to obtain a DETR Data Capture Code for his organisation. This will enable him to:

— create unique reference numbers for his notices using the rules in Appendix E; and

— obtain access to the National Street Gazetteer (NSG) and Additional Street Data (ASD).

D2.2 The undertaker will have to obtain a full set of the DETR Data Capture Codes (available from the DETR website) in order to know what are valid values for each data item on the notice forms.

> **Important Note**: when entering data on the form, undertakers should use the text description of the data, not the code number used by the computer systems. For example, if giving a seven day notice, "SEVEN DAY" should be entered against NOTICE_TYPE not the number "2". This is to minimise errors.

D2.3 The undertaker will need to have an up-to-date NSG and ASD. These are available free of charge to undertakers who have obtained a DETR Data Capture Code, for use solely for street works purposes.

As well as giving the undertaker necessary information (such as about Ttraffic-Ssensitivity), these will enable him to give the correct USRN for the street in his notice.

D2.4 In completing the forms, the undertaker *must* follow the data validation rules set out in Appendix E.

D2.5 The forms themselves are also available in electronic format on the DETR website.

D3 Other Considerations

D3.1 In the event of a fax transmission being not possible because the destination fax machine is engaged, three recorded attempts by the originator within the specified timescale will be deemed to be compliance with the requirements subject to the originator passing the basic information by telephone at that point and confirming it by fax as soon as possible. Where the destination fax machine is unobtainable (rather than engaged), the passing of the basic information by telephone, within the timescale, will be deemed to be compliance subject to its confirmation by fax as soon as possible.

D3.2 The Regulations make provision for authorities and undertakers to provide an address for service of notices under the Act, including (if desired) different addresses for different notices or classes of notice. It is also possible for them to agree other addresses, and means of service. Although an authority or undertaker may wish to specify, for operational reasons, that notices may be sent to local district offices or depots, the number of recipients' addresses for service must be kept to a minimum to avoid confusion and minimise administrative workload.

For information on electronic addresses, see Appendix E3.4.

D.3.3 The precise style and format of the paper notice is discretionary, for example the forms can be adapted to include an organisation's corporate identity or to increase the size of text boxes where they are intended to be handwritten. The layout must however not depart from those defined in respect of the name and order of the data fields. The contents of the text field must always follow the validation rules contained within Appendix E.

It should be noted that, particularly where multiple sites are involved, the form may run over several separate pages.

D4   Relationship of Notices, Works and Sites for charging purposes under The Street Works Register (Registration Fees) Regulations 1999 (S.I.1999/1048).

D4.1   A works may contain one street or more than one street.

D4.2   A works may contain one notice or more than one notice.

D4.3   A works may contain one site or more than one site.

D4.4   A notice can only contain one street unless it is a section 54 notice, in which case it can contain more than one street.

D4.5   A site can contain only one street.

D4.6   The reference of a works (Promoters_Works_Ref) is unique nationally and the Promoters Works Ref will be twenty-four characters in length. It will be constructed as detailed in E3.6.1 of Appendix E.

D4.7   The reference of a site is unique nationally as it consists of the Promoters Works Ref and the Site Num.

D4.8   The reference of a notice is unique nationally as it consists of the Promoters Works Ref and the Unique Street Reference Number (USRN) for a street and the Site Num for that street. If there is more than one street in a works then each notice reference will be based on the information shown above except that the Site Num to be used in each notice will be the first Site Num shown within the works for each particular street.

D4.9   Charging of the receipt and input of non-electronic notifications to the highway authority's street works register of non-electronic notifications is carried out on a street basis, i.e. the charge is made for the input of each individual street (see Figure 1).

D5   Notice Forms

The notice forms at D5.1, D5.2 and D5.3 are to be used for paper notifications. Electronic copies of these forms are will be made available on the DETR Web Site website as soon as possible after going to press.

Figure 1

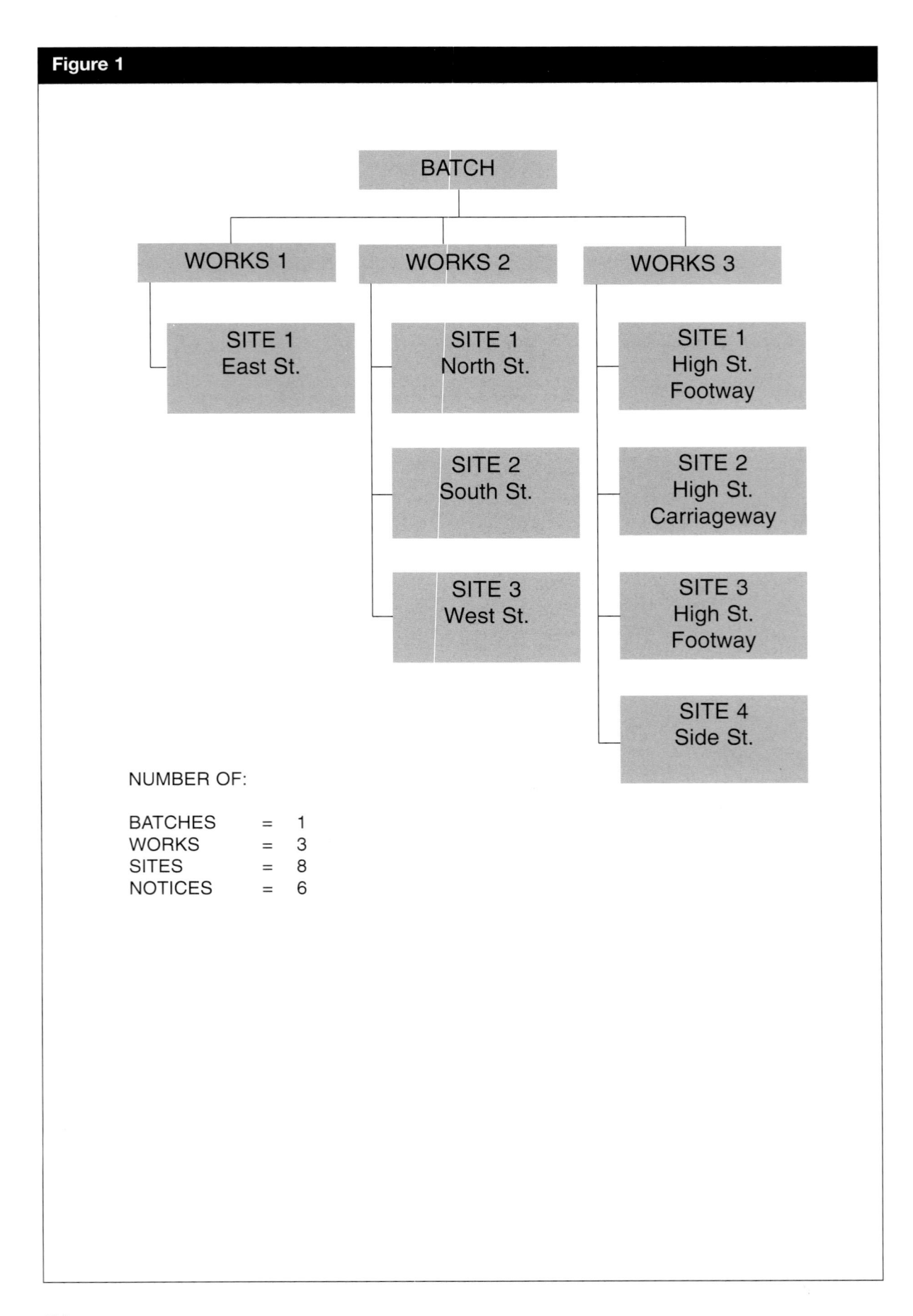

```
                        BATCH
           ┌──────────────┼──────────────┐
        WORKS 1        WORKS 2        WORKS 3
           │              │              │
        SITE 1         SITE 1         SITE 1
        East St.       North St.      High St.
                                      Footway

                       SITE 2         SITE 2
                       South St.      High St.
                                      Carriageway

                       SITE 3         SITE 3
                       West St.       High St.
                                      Footway

                                      SITE 4
                                      Side St.
```

NUMBER OF:

BATCHES   = 1
WORKS   = 3
SITES   = 8
NOTICES   = 6

Appendix D — Paper Transfer of Notices

D5.1 Works Form

FROM:	
TO:	

	WORKS
UTILITY REFERENCE:	
VERSION NUMBER:	
CREATED:	
ABANDONED:	
NOTICE TYPE:	
WORKS TYPE:	
WORKS STATUS:	
DESCRIPTION:	
INSPECTION UNITS & TYPE:	
START DATE & TIME:	
ESTIMATED END DATE:	
COMPLETED DATE:	
CONTACT (OWNER):	
CONTACT (AGENT):	

	SITE
SITE NO.:	
VERSION:	
STREET USRN:	
STREET NAME:	
LOCALITY:	
TOWN:	
COUNTY:	
CREATED:	
EXTANT DATE:	
COMPLETION DATE:	
ABANDONED DATE:	
SITE STATUS:	
TRAFFIC MANAGEMENT:	
LOCATION:	
POSITION:	
EXCAVATION DEPTH:	
LENGTH:	
WIDTH:	

	PROVISIONAL STREET
STREET NAME:	
AREA:	
TOWN:	
COUNTY:	
OSGRs:	

SITE COORDINATES		
NUMBER	EASTING	NORTHING

SPECIAL DESIGNATIONS

D5.2 Comments Form

FROM:	
TO:	

COMMENTS	
PROMOTER WORKS REF:	
SITE NUM:	
COMMENT NUMBER:	
COMMENT TYPE:	
COMMENT TEXT:	
COMMENT DATE:	
NSG STREET IDENTIFIER:	

D5.3 Inspections Form

Inspections form will be contained in the Inspections Code of Practice for Inspections.

APPENDIX E
Electronic Transfer of Notices

E1 Introduction

E1.1 This Appendix sets out a methodology for the electronic transfer of notices between undertakers and street authorities.

E1.2 The notices dealt with in this Appendix are those required by sections 54, 55, 57, 58, 70, 74, 81 and 85 of the New Roads and Street Works Act 1991 and also non-statutory notifications currently sent by means of Daily Whereabouts. Other types of notice may be added later. In the meantime, other types of notice should be given as previously.

E1.3 Notices may continue to be given by methods other than those set out in this Appendix. However where a highway authority has the means to receive the notices required by sections 54, 55, 57 and 70 of the Act and Daily Whereabouts electronically, non-electronic notices can attract a registration fee (see the Street Works Register (Registration Fees) Regulations 1999, S.I.1999/1048). Consideration is being given to the introduction of registration fees for non-electronic transmission of other types of notice also, but any decision to do this would require consultation and further regulations.

E1.4 This version of Appendix E is denoted as version 2.0 within the batch header.

E2 Electronic Transfer of Notices

E2.1 With effect from 1 April 1999, highway authorities must ensure that they have in place the means to receive electronically notices given in the form defined in section E3 and sent by the method defined in section E4.

E2.2 With effect from 1 April 1999, highway authorities must ensure that they retain the capability to receive notices by fax, post or delivery. However, notices received in these fashions will attract a fee as stated in Regulations.

E2.3 In the event of an electronic transmission not being possible because the receiving server is engaged or unavailable, three recorded attempts by the originator will be deemed to be compliance with the requirements subject to the originator passing the basic information by telephone at that point and confirming it by electronic transmission as soon as possible.

E2.4 With effect from 1 April 2001, undertakers are strongly recommended to ensure that they have in place the means to receive electronically notices given in the form defined in section E3 and sent by the method defined in section E4.

E3 Data Formats

E3.1 Formats for Data Exchange

This section defines data exchange formats and the mechanisms that are to be used with Street Works Registers. The formats are designed for different Street Works systems to work together.

The formats describe the following data:

Organisation and Operational Districts.
Areas of interest (of Operational Districts)
} Allowing Organisations to determine who relates to their own area.

Street Works Batches — Transferred in accordance with the Notification rules as follows

— to street authorities: sections 54, 55, 57, 70(3), 70 (registrations), 74 and Daily Whereabouts, and

— to undertakers: sections 58, 74 and 85.

Works Comments, including directions under sections 56 and 65, and notices under sections 66 and 81.

Each data format will be preceded by a description of the relationship between the various data attributes. All coded values are contained in the DETR's Data Capture Codes which are obtainable from the DETR website.

Inspections including defects and the details of the sample/random inspections will be found in the Inspections Code of Practice.

E3.2 Batch Structure

Data may be batched according to the following rules. Batches may not be amalgamated into bigger files.

Each batch of data will contain only one batch header providing information on the data contained within it as well as the reference of the Organisation that created it (i.e. SWA_Org_Ref and District_Ref). A batch will contain only one type of data (e.g. Street Works).

A batch line can be up to 510 characters followed by a line terminator (the line terminator would be a carriage return and line feed character on a PC or a carriage return on a UNIX system). Blank lines can be used to divide the batch but tab characters must not be used (use spaces for indentation). Keywords (which are shown in this document in capitals) must be separated from data by at least a single space. In a batch transfer, keywords must be present and be in capitals. All other data items are not case sensitive.

Where 'null' is specified for a data item within the data validation table, that data item should consist of the colon (:) followed immediately by a line terminator. 'Space' characters or zero values should not be used.

The repeating group indicators [and] (square brackets) must always be present. Where a repeating group contains no information then the 'start of repeating group' can be followed on the next line by the 'end of repeating group' (i.e. no information has to be supplied even if there are internal repeating groups). For example, where a works batch only updates Works information and says nothing about any of the Sites, the end of the batch would look as follows:

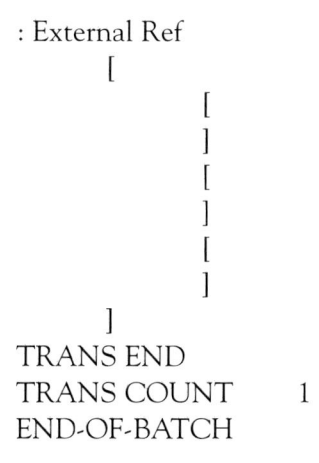

```
: External Ref
        [
                [
                ]
                [
                ]
                [
                ]
        ]
TRANS END
TRANS COUNT      1
END-OF-BATCH
```

All the attributes must be supplied and presented in the defined order. Optional values that are not provided will be shown by a line with only a colon (:). Lines where the first non-space character is an exclamation mark (!) are comments. There are no limits to the number of comment lines.

Examples of 'Batch Headers' are contained in the data definitions below, a typical batch header is:

```
HEADER_BEGIN
        USER_BATCH_ID              28
        BATCH_PRIORITY_CODE        S
        VERSION                    2.0
        SWA_ORG_REF                1100
        DISTRICT_REF               001
HEADER_END
```

Item	Optional or Mandatory	Description
HEADER_BEGIN	M	The first record in the batch.
USER_BATCH_ID	M	A reference of the batch creator. (Maximum 50 characters)
BATCH_PRIORITY_CODE	M	The priority of the Batch: S — Process sequentially in the order the batches are received. A — Process as soon as possible (for the creation of emergency and urgent works only).
VERSION	M	Version of batch definitions in range 1.0 to 999.9 (This is version 2.0)
SWA_ORG_REF	M	DETR Data Capture code.
DISTRICT_REF	M	Created by Organisation with a range of 1 to 999. This is for the sending organisation's reference only and the receiving organisation should not attribute any significance to it.
HEADER_END	M	End of batch header details.

The 'Batch Header' is followed by the 'Transactions' (Data) to be processed. Each 'Transaction' has the following structure:

```
TRANS_BEGIN        works_batch
TRANS_REF          1100_001:97J03124
!Comment
:Data
     [
     :More Data — Repeating Group
          [
          :Even More Data — Repeating Group
          ]
     ]
TRANS_END
```

Item	Optional or Mandatory	Description
TRANS_BEGIN	M	DETR Data Capture code (e.g. works_batch).
TRANS_REF	O	A reference of the batch creator. (Maximum 50 characters)
!	O	A comment.
:Data	O/M	Data items always start with a colon. Space before the colon is ignored.
[	M	Start of repeating data items.
:More Data	O/M	The repeating data items.
[	M	Start of repeating data items.
:Even More Data	O/M	The repeating data items.
]	M	End of repeating data items.
]	M	End of repeating data items.
TRANS_END	M	The end of all the data for the particular 'Transaction'.

After all the 'Transactions' in a batch there is:

TRANS_COUNT   1
END_OF_BATCH

Item	Optional Mandatory	Description
TRANS_COUNT	M	The number of 'Transactions' in the batch.
END_OF_BATCH	M	The last record in the batch.

(Note: The indenting has been used to add clarity when viewing a batch).

A batch may contain more than one 'Transaction' providing they are all of the same type.

E3.2.1 BATCH DATA

The definition of the minimum set of data to be exchanged is described in the following sections. A table follows each data definition detailing the data items. Where a 'Mandatory' item is followed by an asterisk (i.e. M*), this indicates that the attribute is mandatory when the record is created but is 'Optional' when updates occur.

Each type of batched data is given a description on how it is used in various circumstances (i.e. new or updated records). The significance of some data items is also described (e.g. versioning).

E3.3 Streets

Streets as defined by Level I of the National Street Gazetteer (NSG) and Additional Street Data are available from the NSG Concessionaire. The data definition and transfer formats are defined in Section E6 of this document.

E3.3.1 RELEVANT AND INTERESTED AUTHORITIES

"Relevant Authorities" are defined in section 49(6) of the Act. These are sewer, bridge or transport authorities according to circumstances. Relevant authorities are entitled to copies of section 55 notices. In addition, to facilitate co-ordination, some street authorities wish to receive copies of notices relating to works in streets adjacent to their area boundaries or streets within their boundaries, which have a different street authority (e.g. the Highways Agency).

This is facilitated by creating a Type 21 record for the street authority (indicated by setting the SWA_ORG_TYPE to "1") and further Type 21 records for other interested authorities (with the SWA_ORG_TYPE set to the relevant value, other than "1").

The Operational Districts of the relevant or interested authorities will receive all street works and site information. However a batch may only contain information relating to only one Operational District.

E3.4 Operational District Batch Structure

E3.4.1 OPERATIONAL DISTRICTS

The Operational Districts data will be transferred on a complete replacement basis. The highway authority Operational District batch files are available on the NSG website. National undertakers must send a copy of their Operational District batch files to all highway authorities; regional undertakers must send a copy of their Operational District batch files to all highway authorities in the region in which they operate.

E3.4.2 TCP/IP ADDRESS AND DOMAIN NAME

The address of the computer that will receive data maybe given as either the TCP/IP address or the Domain Name.

E3.4.3 DIRECTORY ADDRESS

The directory of the computer that is to receive data.

E3.4.4 OPERATIONAL DISTRICTS DATA DEFINITION

```
HEADER_BEGIN
        USER_BATCH_ID              30
        BATCH_PRIORITY_CODE        S
        VERSION                    2.0
        SWA_ORG_REF                1100
        DISTRICT_REF               001
HEADER_END
TRANS_BEGIN        od_batch
TRANS_REF          1100_001:97J03144.127
:SWA_Org_Ref
:SWA_Org_Name_Text
        [
        !Districts
        :District_Ref
        :District_Name_Text
        :District_Address_Text
        :District_Post_Code_Text
        :District_Tel_No_Text
        :District_Closed
        :Operational_District_Code
        :TCP_IP_Addr
        :Domain_Addr
        :Dir_Addr
        ]
TRANS_END
TRANS_COUNT        1
END_OF_BATCH
```

E3.4.5 OPERATIONAL DISTRICTS DATA VALIDATION

Attribute	Optional or Mandatory	Validation
SWA_Org_Ref	M	DETR Data Capture code.
SWA_Org_Name_Text Districts	M	Up to forty characters.
District_Ref	M	Created by Organisation with a range of 1 to 999.
District_Name_Text	M	Up to forty characters.
District_Address_Text	M	Up to two hundred characters.
District_Post_Code_Text	M	Up to eight characters and includes spaces.
District_Tel_No_Text	M	Up to twenty five characters.
District_Closed	O	YYYY-MM-DD The date on which the district was closed.
Operational_District_Code	M	5 character works reference prefix see E3.6.1.
TCP_IP_Addr	O	TCP/IP.
Domain_Addr	O/M	Up to one hundred characters. Mandatory if TCP/IP address not completed, does not need to include the "ftp://" prefix
Dir_Addr	M	Directory address (Up to 50 characters).

E3.5 Areas of Interest Batch Structure

E3.5.1 AREAS OF INTEREST

An Operational District may have many Areas of Interest. The Area of Interest whose number is 1 will define the area enclosing the Operational District. Records are transferred on a complete replacement basis. National undertakers must send a copy of their Areas of Interest batch files to all highway authorities; regional undertakers must send a copy of their Areas of Interest batch files to all highway authorities in the region in which they operate.

E3.5.2 AREAS OF INTEREST DATA DEFINITION

```
HEADER_BEGIN
        USER_BATCH_ID                   31
        BATCH_PRIORITY_CODE             S
        VERSION                         2.0
        SWA_ORG_REF                     1100
        DISTRICT_REF                    001
HEADER_END
TRANS_BEGIN       aoi_batch
TRANS_REF         1100_001:97J03154.127
 :SWA_Org_Ref
        [
        !Districts
        :District_Ref
        :AOI_Polygon_Num
        :AOI_Polygon_Description_Text
                [
                !Coordinates
                :Coordinate_Num
                :Spatial_Ref_Code_E
                :Spatial_Ref_Code_N
                ]
        ]
TRANS_END
TRANS_COUNT       1
END_OF_BATCH
```

E3.5.3 AREAS OF INTEREST DATA VALIDATION

Attribute	Optional or Mandatory	Validation
SWA_Org_Ref	M	DETR Data Capture code.
Districts		
District_Ref	M	Created by Organisation with a range of 1 to 999.
AOI_Polygon_Num	M	Polygon number.
AOI_Polygon_Description_Text	M	Up to one hundred and twenty characters.
Coordinates		
Coordinate_Num	M	Up to 9999 points (the minimum number of points must not be less than 4 and the last point must be the same as the first). The points should be sequentially numbered without gaps.
Spatial_Ref_Code_E	M	A 6 figure Easting to one metre precision.
Spatial_Ref_Code_N	M	A 6 figure Northing to one metre precision.

E3.5.4 INTER-UTILITY EXCHANGE OF NOTICES

Areas of Interest may be used to facilitate inter-utility exchange of notices.

Areas of Interest data must be provided by the undertaker to all of the other undertakers from whom copy notices are required. This will be used by the sending undertaker to determine whether the coordinates of either end points of the street are within the defined polygon. The copy notice is submitted using the address details within the relevant Operational District Batch File. Undertaker Areas of Interest and Operational District batch files are available from the NJUG website.

E3.6 Street Works Batch Structure

E3.6.1 STREET WORKS

The principle that the reference of all Street Works will be unique nationally will be implemented. The Works Promoters works reference will be up to 24 characters in length. The first 5 characters will be the Operational_District_Code constructed as follows:

The first two characters are issued by the DETR (Department of Environment, Transport and the Regions) for the Organisation.

The next three characters are the District_Ref, right justified with leading zeros.

The last nineteen characters are whatever the Operational District requires.

The reference must not include single (') and double quotes (").

The Rules in D4 about works and sites also apply to electronic notices.

E3.6.2 NATIONAL STREET GAZETTEER SELECTION FOR SITES

Street Works Sites may only be located against National Street Gazetteer references of Type 1 (Designated Name) or Type 2 (Street Description). Sites allocated against Type 3 (Road Number) and Type 4 (Unofficial Name) are not permitted.

E3.6.3 PROVISIONAL STREETS

Where a Works Promoter is unable to find a street in the gazetteer every effort should be made to ensure that it genuinely does not exist (i.e. contact should be made with the Highway Authority responsible for the gazetteer). If a provisional street has to be created then the NSG_Street_Identifier_Ref used by the Works Promoter in the Street Works Site definition should be zero with the Street, Area, Town, County, Easting and Northing defined accordingly.

E3.6.4 WORKS AND SITES VERSIONS

A Works or Site may be 'Versioned' where the sending Organisation decides a changed item of data is of historical interest (e.g. the Works Status has changed from 'Proposed' to 'Permanent Complete'). Where more than one version is sent in a batch then they should be in 'Version Number' order (i.e. all the Works Versions should precede all the Site Versions). There is no requirement to link Site Versions to Works Versions.

E3.6.5 NOTICE AND WORKS TYPES

The 'Notice Type' will be determined by the sending organisation. It will be the sending Organisation's responsibility to adhere to the prescribed periods (either Notice or Daily Whereabouts).

E3.6.6 EXTERNAL REFERENCE

Where the Street Works is part of an Organisation's larger project that may not all be on the highway there is sometimes a need for a link from the Street Works to the larger project. The 'External Reference' is the reference of the larger project.

E3.6.7 SECTION 74

Note that, as a general rule all works batches must contain at least one site. However the introduction of section 74 procedures has required some changes to the procedures.

Where a section 74 notice of Works Clear or Works Closed is given and that notice is not intended to also include either a section 70(3) notice or a registration of reinstatement, then that works batch must NOT include any sites.

Where a section 74 notice of Works Clear or Works Closed is given and that notice is intended to include a section 70(3) notice, then that works batch MUST include the relevant sites.

Where a section 74 notice of Works Clear or Works Closed is given and that notice is intended to also include a registration of reinstatement, then that works batch MUST include the relevant sites' length, width and depth dimensions. This will necessarily also stand as a section 70(3) notice if such notice has not previously been given.

E3.6.8 STREET WORKS DATA DEFINITION

```
HEADER_BEGIN
        USER_BATCH_ID              32
        BATCH_PRIORITY_CODE        S
        VERSION                    2.0
        SWA_ORG_REF                1100
        DISTRICT_REF               001
HEADER_END
TRANS_BEGIN        works_batch
TRANS_REF          1100_001:97J03164.127
:Promoter_Works_Ref
:Works_Version_Num
:Works_Version_Created_Datim
:Works_Abandoned_Datim
:Notice_Type
:Works_Type_Code
:Works_Status_Code
:Works_Description_Text
:Works_Insp_Units_Num
:Works_Insp_Units_Type_Code
:Works_Start_Date
:Works_Start_Time
:Works_Estimated_End_Date
:Works_Completed_Date
:Works_Contact_Name_Owner
:Works_Contact_Address_Owner
:Works_Contact_Tel_No_Owner
:Works_Contact_PostCode_Owner
:Works_Contact_Name_Agent
:Works_Contact_Address_Agent
:Works_Contact_Tel_No_Agent
:Works_Contact_PostCode_Agent
:Restricted_Access_Flag
```

```
            :External_Ref
                    [
                    !Sites
                    :Site_Num
                    :Site_Version_Num
                    :Site_Version_Created_Datim
                    :NSG_Street_Identifier_Ref
                    :Site_Extant_Date
                    :Site_Proposed_End_Date
                    :Site_Abandoned_Date
                    :Site_Status_Code
                    :Site_Spatial_Coord_Type_Code
                    :Site_Traffic_Management_Code
:Street_Location_Code
                    :Site_Location_Text
                    :Site_Location_Post_Code_Text
                    :Site_Depth_Code
                    :Site_Length_Val
                    :Site_Width_Val
                            [
                            !Provisional Street
                            :Street
                            :Area
                            :Town
                            :County
                            :Street_East
                            :Street_North
                            ]
                            [
                            !Site Coordinates
                            :Coordinate_Num
                            :Spatial_Ref_Code_E
                            :Spatial_Ref_Code_N
                            ]
                            [
                            !Site in Special Designations
                            :Street_Special_Desig_Code
                            ]
                    ]
        TRANS_END
        TRANS_COUNT     1
        END_OF_BATCH
```

E3.6.9 STREET WORKS DATA VALIDATION

Attribute	Optional or Mandatory	Validation
Promoter_Works_Ref	M	Constructed with the unique five character Operational_District_Code and up to nineteen additional characters. (In the case of a section 58 or a section 85 notice the 'promoter' would be the Street Authority).
Works_Version_Num	M	The Version of the Works record being sent (Must be zero if Versioning is not used).
Works_Version_Created_Datim	M	YYYY-MM-DD:HH:MM:SS.SS The time the Works Version was created or last modified and must include two decimal places for seconds (Which may be 00).
Works_Abandoned_Datim	M	YYYY-MM-DD Mandatory to give date if works are abandoned otherwise null
Notice_Type	M	DETR Data Capture code.
Works_Type_Code	M*	DETR Data Capture code.
Works_Status_Code	M*	DETR Data Capture code. The Works Status is the status of the whole of the works, not just the status of those sites in any particular batch.
Works_Description_Text	M*	Up to five hundred characters. (This should be written in plain English without jargon or abbreviations not understood outside the sending organisation.)
Works_Insp_Units_Num	M*	Number of inspection units (can be zero). Null for Street Authority notices issued under sections 58 and 85.
Works_Insp_Units_Type_Code	M*	DETR Data Capture code. Null for Street Authority notices issued under sections 58 and 85.
Works_Start_Date	M*	YYYY-MM-DD.
Works_Start_Time	M	HH:MM:SS Mandatory for Emergency, Special Urgent and Urgent Works otherwise null.
Works_Estimated_End_Date	M*	YYYY-MM-DD
Works_Completed_Date	M	YYYY-MM-DD Mandatory if Works is Interim Complete, Permanent Complete, Clear or Closed unless the date has not changed from the last time a works completed date was given otherwise null. This is the same as the latest Site_Extant_Date of sites which are at the same status as the works.
Works_Contact_Name_Owner	O/M	Up to forty characters (Mandatory if Works is 'Under Licence' or 'Statutory Right' for a private developer).

Attribute	Optional or Mandatory	Validation
Works_Contact_Address_Owner	O/M	Up to two hundred characters (Mandatory if Works is 'Under Licence' or 'Statutory Right' for a private developer).
Works_Contact_Tel_No_Owner	O	Up to twenty five characters.
Works_Contact_PostCode_Owner	O	Up to eight characters and includes spaces.
Works_Contact_Name_Agent	O	Up to forty characters.
Works_Contact_Address_Agent	O	Up to two hundred characters.
Works_Contact_Tel_No_Agent	O	Up to twenty five characters.
Works_Contact_PostCode_Agent	O	Up to eight characters and includes spaces.
Restricted_Access_Flag	O	A flag to prevent unauthorised organisations viewing this Works. Set to 1 for restricted access.
External_Ref	O	The link to an Organisations internal systems (Up to 20 characters).
Sites		
Site_Num	M	Up to nine hundred and ninety nine Sites.
Site_Version_Num	M	The Version of the Site being sent (Must be zero if Versioning is not used).
Site_Version_Created_Datim	M	YYYY-MM-DD:HH:MM:SS.SS The time the Site Version was created or last modified and must include two decimal places for seconds (Which may be 00).
NSG_Street_Identifier_Ref	M	A NSG Street Reference which must be of 'Reference Type' 1 or 2. This will be zero if the Site is on a 'Provisional Street'.
Site_Extant_Date	M	YYYY-MM-DD. The date on which the activity is expected to start on the Site or the date on which the activity was completed or the date on which the site was abandoned.
Site_Proposed_End_Date	M	YYYY-MM-DD. The date on which the activity is expected to finish on any proposed Site otherwise null.
Site_Abandoned_Date	M	YYYY-MM-DD. The date on which the Site was abandoned (i.e. no activity took place whatsoever) otherwise null.
Site_Status_Code	M	DETR Data Capture code.
Site_Spatial_Coord_Type_Code	O	DETR Data Capture code.
Site_Traffic_Management_Code	O	DETR Data Capture code. Null for Street Authority notices issued under sections 58 and 85.
Street_Location_Code	O/M	DETR Data Capture code. (Optional for Proposed Sites).
Site_Location_Text	M*	Up to one hundred and twenty characters. Must be consistent with the USRN for this site. Site_Location_Post_Code_Text O Up to eight characters and includes spaces.

Appendix E — Electronic Transfer of Notices

Attribute	Optional or Mandatory	Validation
Site_Depth_Code	M	DETR Data Capture code. Mandatory when the Site_Status is Interim Reinstatement or Permanent Reinstatement or Remedial Reinstatement which re-sets the guarantee period AND the Notice_Type is NOT section 70; optional when the works status is Clear or Closed; in all other cases, including Street Authority notices issued under sections 58 and 85, the field must be null.
Site_Length_Val	M	Up to 9999.99 metres Mandatory when the Site_Status is Interim Reinstatement or Permanent Reinstatement or Remedial Reinstatement which re-sets the guarantee period AND the Notice_Type is NOT section 70; optional when the works status is Clear or Closed; in all other cases, including Street Authority notices issued under sections 58 and 85, the field must be null.
Site_Width_Val	M	Up to 99.99 metres Mandatory when the Site_Status is Interim Reinstatement or Permanent Reinstatement or Remedial Reinstatement which re-sets the guarantee period AND the Notice_Type is NOT section 70; optional when the works status is Clear or Closed; in all other cases, including Street Authority notices issued under sections 58 and 85, the field must be null.
Provisional Street (Repeating group brackets are supplied so that nothing can be entered)		
Street	M	'Provisional Street' name.
Area	O	'Provisional Street' Area Name.
Town	M	'Provisional Street' Town Name.
County	M	'Provisional Street' County Name.
Street_East	M	A 6- figure Easting to 1 metre precision on a 'Provisional Street'.
Street_North	M	A 6-figure Northing to 1 metre precision on a 'Provisional Street'.
Site Coordinates		
Coordinate_Num	O	The next Site coordinate.

Attribute	Optional or Mandatory	Validation
Spatial_Ref_Code_E	O	A 6 figure Easting to 1 metre precision should be given where practicable and in rural areas where no other unique identification is available.
Spatial_Ref_Code_N	O	A 6 figure Northing to 1 metre precision should be given where practicable and in rural areas where no other unique identification is available.
Site in Special Designations Street_Special_Desig_Code	O	The DETR Data Capture code where it applies to the activity on this site. To be left blank for section 58 and section 85 notices.

E3.7 Works Comments Batch Data Structure

E3.7.1 WORKS COMMENTS

Directions under sections 56 and 65 and notices under sections 66, 74 and 81 are transferred in this data.

Works comments must be uniquely numbered and, in particular, any amendment to a comment already sent must be sent as a new, uniquely numbered comment and not re-sent using the original number.

E3.7.2 WORKS COMMENTS DATA DEFINITION

```
HEADER_BEGIN
    USER_BATCH_ID            33
    BATCH_PRIORITY_CODE      S
    VERSION                  2.0
    SWA_ORG_REF              1100
    DISTRICT_REF             001
HEADER_END
TRANS_BEGIN        comments_batch
TRANS_REF          1100_001:97J03174.127
:Promoter_Works_Ref
    [
    : Site_Num
    :Comment_Num
    :Comment_Type_Code
    :Comment_Text
    :Comment_Made_Datim
    :NSG_Street_Identifier_Ref
    :Relevant_Date
    ]
```

```
TRANS_END
TRANS_COUNT    1
END_OF_BATCH
```

E3.7.3 WORKS COMMENTS DATA VALIDATION

Attribute	Optional or Mandatory	Validation
Promoter_Works_Ref	M	The 'Works Reference' for the Comment. (See E3.7.4)
Site_Num	M	Mandatory where the Comment is Site specific. Null where the Comment applies to whole Works.
Comment_Num	M	The unique sequential number of the Comment from the 'Commenting' Organisation.
Comment_Type_Code	M	DETR Data Capture code.
Comment_Text	M	Up to five hundred characters.
Comment_Made_Datim	M	YYYY-MM-DD:HH:MM:SS.SS (Must include 2 decimal places for seconds).
NSG_Street_Identifier_Ref	M	NSG Street Reference (USRN) for Site Specific Comments and section 81 comment. Null for Works Comments.
Relevant_Date	M	YYYY-MM-DD Must be completed with the revised completion date under section 74 challenge otherwise null.

E3.7.4 WORKS COMMENTS INFORMATION

Where Works Comments are for proposed or existing works the Promoter_Works_Ref is the Undertakers Promoter_Works_Ref on their original works. In the case of section 81 notices, or where there are unattributable works i.e. works with no notice, the highway authority will construct the Promoter_Works_Ref using their own reference and their own two DETR characters all in accordance with Section E3.6.1. Where an undertaker accepts responsibility for previously unattributable works he must issue relevant notices using his own Promoter_Works_Ref, not that generated by the highway authority.

E3.8 Inspection Data Batch Structure

E3.8.1 INSPECTIONS

The inspection details to be transferred will include:

— defective inspections.

— the details of the sample/random inspections (equivalent to the inspector's check sheet).

E3.8.2 INSPECTIONS DATA DEFINITION

The inspections data definition will be found in the Inspections Code of Practice.

E3.8.3 INSPECTIONS DATA VALIDATION

The inspections data validation will be found in the Inspections Code of Practice.

E4 Data Transfer Method

E4.1 The applications programmes will use FTP (File Transfer Protocol) to transfer data between systems. FTP has to be communicated using the TCP/IP protocols (Transmission Control Protocol/Internet Protocol) used by internet computers. The character set and file format must be ASCII. Undertakers and street authorities will make known their relevant internet address or domain address to each other. Recipients of electronic notice information must use an open and unrestricted FTP Server with anonymous access.

As this method is a direct connection between computers, the file naming convention for exchanged batches will be XOOOOdddYYYYMMDDhhmmss.txt where X is the Batch_Priority_Code, OOOO is the DETR Capture Code for each organisation (SWA_ORG_REF), ddd is the District_Ref, YYYY is the year, MM is the month, DD is the day (1st=01), hh is the hour, mm the minute and ss the seconds which represents the date and time the batch was created and will assist the receiving Organisation's processing sequence. All numeric fields must be zero-filled. All text fields are case sensitive and must be in the case shown.

The return path from Street Authority to Undertakers is derived from the Promoter_Works_Ref.

E5 Data Capture Document

E5.1 This document is available from the Department of the Environment, Transport and the Regions on their Internet site at URL http://www.detr.gov.uk.

E6 Street Data

This section defines the data exchange formats for the NSG and Additional Street data.

The formats described are those used for transfer to and from the NSG concessionaire and some of the data fields may not be required for use with Electronic Transfer of Notices. The data formats are the same as those defined within BS7666 and associated guidance documentation however some of the field names may vary between the British Standard and the formats defined within this document. The definition, structure, and validation rules for the fields remain the same and therefore the different field names have no affect on the data being transferred.

E6.1 Formats for Data Exchange

The Street Authorities will submit their data to the NSG Concessionaire, which will be provided to Undertakers via the Internet in an ASCII-text, comma separated value transfer set (CSV).

File naming should follow the form "*xxxx_nn*.CSV" where *xxxx* is the appropriate SWA_ORG_REF for the Authority producing the information and *nn* is;

(a) for gazetteer data, a sequential number allocated by an authority where it has produced more than one local gazetteer for its area, and

(b) for associated street data, the record type number (21, 22 or 23).

E6.1.1 RECORD STRUCTURE

The transfer set will contain a number of different record types, one for each of the different NSG records. The first field in each of these records will be a Record Identifier, which will determine the content and format of the remainder of the physical record. There will be one record per line in each file and there will be carriage return, line feed characters at the end of each line.

All files will contain HEADER and TRAILER records as the first and last records in the file. The order of all other records within each file is unimportant.

In each of the records, the data items (fields) listed in this specification will be included in the order that they occur in the relevant record definition. Each field shall must be separated from the previous by a comma. Text fields must be delimited, i.e. contained within double quotation marks.

E6.2 Record Structures – Administrative Records

The following sections describe the structure of those administrative records used to control the transfer of street data. These records are implemented purely to allow the consistent transfer of street data, and are not a part of the data itself.

E6.2.1 HEADER record (10)

The HEADER record is compulsory for all transfer sets. A HEADER record will be the first record in every volume.

Field	Description	Type	Maximum Length	Value Range	Mandatory?
RECORD_IDENTIFIER	Identifies this record as a HEADER record.	numeric	2	10	Yes
SWA_ORG_NAME_TEXT	Name of the organisation providing the data.	string	40		Yes
SWA_ORG_REF	User organisation DETR code.	numeric	4	DETR Data Capture Code	Yes
PROCESS_DATE	Process date. The date when the NSG CSV transfer set was created. This date will be the same for all volumes in the transfer set.	date	10	1998-04-01 to present date	Yes
VOLUME_NUMBER	The Volume Sequence Number	numeric	2	01-99	Yes

Record Example

```
10,"DEVON COUNTY COUNCIL",1100,1998-04-09,
```

E6.2.2 TRAILER record (99)

The TRAILER record is compulsory for all transfer sets. A TRAILER record will be the last record in every volume.

Field	Description	Type	Maximum Length	Value Range	Mandatory?
RECORD_IDENTIFIER	Identifies this record as a TRAILER record.	numeric	2	99	Yes
NEXT_VOLUME_NUMBER	The Next Volume Sequential Number	numeric	2	01-99	Yes
RECORD_COUNT	Count of the number of records in the volume (excluding the HEADER and TRAILER records)	numeric	14		Yes
CHECK_SUM	The Automatic Check Sum Value	string	8		No

Record Example

```
99,,15674,
```

E6.3 Record Structures – NSG Records

The following sections describe the structure of each of the records used for the transfer of NSG data.

E6.3.1 STREET record (11)

Field	Description	Type	Maximum Length	Value Range	Mandatory?
RECORD_IDENTIFIER	Identifies this record as a STREET record.	numeric	2	11	Yes
USRN	Unique street reference number.	numeric	8		Yes
RECORD_TYPE	Street reference type.	numeric	1	1-4 Defined in BS7666	Yes
STREET_DESCRIPTOR	Name, description or street number.	string	100		Yes
LOCALITY_NAME	Locality name.	string	35		No
TOWN_NAME	Town name.	string	30		No
COUNTY_NAME	County name.	string	30		Yes
ALIAS_STREET_DESCRIPTOR	Name, description or street number alias for the street.	string	100		No
ALIAS_LOCALITY_NAME	Locality name alias for the street.	string	35		No
ALIAS_TOWN_NAME	Town name alias for	string	30		No
ALIAS_COUNTY_NAME	County name alias for the street.	string	30		No
SWA_ORG_REF_NAMING	The DETR code of the Street Naming Authority if named or described, or the Highway Authority if numbered.	numeric	4	DETR Data Capture Code	Yes
STREET_VERSION_NUMBER	A sequential number indicating the version of record.	numeric	4	1-9999	Yes
STREET_ENTRY_DATE	The date on which the record was entered or a new version created.	date	10	1990-01-01 to present date +1 year	Yes

Field	Description	Type	Maximum Length	Value Range	Mandatory?
STREET_CLOSURE_DATE	The date on which the street was closed or a new version replaced the record.	date	10		No
STREET_START_X	The X (eastings) co-ordinate of the start point of the street. Co-ordinates are defined in metres.	numeric	7	0-660000	Yes
STREET_START_Y	The Y (northings) co-ordinate of the start point of the street. Co-ordinates are defined in metres.	numeric	7	0-1300000	Yes
STREET_END_X	The X (eastings) co-ordinate of the end point of the street. Co-ordinates are defined in metres.	numeric	7	0-660000	Yes
STREET_END_Y	The Y (northings) co-ordinate of the end point of the street. Co-ordinates are defined in metres.	numeric	7	0-1300000	Yes
STREET_TOLERANCE	The tolerance of the start and end co-ordinates. Tolerance is defined in metres.	numeric	3	0-999	Yes

Record Example

```
11,12345678,1,"ANY STREET",,"ANYTOWN",,,,,,1234,1,1998-04-09,,123456,234567,345678,456789,10
```

E6.3.2 STREET_XREF record (12)

Field	Description	Type	Maximum Length	Value Range	Mandatory?
RECORD_IDENTIFIER	Identifies this record as a STREET_XREF record.	Numeric	2	12	Yes
XREF_TYPE	Indicator as to type of record that is cross-referenced. This is an explicit cross reference and is always one.	Numeric	1	1	Yes
USRN	Unique street reference number.	Numeric	8		Yes
USRN_VERSION_NUMBER	A sequential number indicating the version of the street for which the cross-reference applies.	Numeric	4	1-9999	Yes
XREF_ID	Cross references to other representations of the street.	numeric	14		Yes
XREF_VERSION_NUMBER	A sequential number indicating the version of the street that is being cross-referenced to.	Numeric	4	1-9999	Yes

Record Example

```
12,1,12345678,5,23456789,3
```

E6.4 Record Structures – Additional Records

The following sections describe the structure of each of the records used for the transfer of additional data in support of the NSG.

These additional data records may be transferred in any combination, within the restrictions defined for each record.

E6.4.1 ADDITIONAL_STREET record (21)

The Additional Street record is used to supply additional street attribute data for the NSG. This information is only supplied for type 1 and 2 streets (i.e. designated street names and street descriptions).

Field	Description	Type	Maximum Length	Value Range	Mandatory?
RECORD_IDENTIFIER	Identifies this record as an ADDITIONAL_STREET record.	Numeric	2	21	Yes
USRN	Unique street reference number.	Numeric	8		Yes
ADDITIONAL_STREET_SEQ_NUM	Sequential number for each street for each additional street information record.	Numeric	3		Yes
SWA_ORG_REF_AUTHORITY	A code for the authority having an interest in the street.	numeric	4	DETR Data Capture Codes	Yes
WHOLE_ROAD	Indicator as to whether the additional street information applies to the whole road. 0 indicates that it does not apply to the whole road, 1 indicates that it does.	numeric	1	0,1	Yes
ADDITIONAL_STREET_LOCATION_TEXT	Description of the location of the part(s) of the street for which this additional street record is applicable	string	120		Yes unless WHOLE_ROAD indicator = 1
DISTRICT_REF_AUTHORITY	The code for the Operational District within the authority.	numeric	3	District_Ref	Yes (E3.4.5)
SWA_ORG_REF_MAINTAINING_DATA	A code for the organisation responsible for maintaining the data on the street.	numeric	4	DETR Data Capture Codes	No
DISTRICT_REF_MAINTAINING_DATA	The code of the Operational District within the maintaining authority responsible for maintaining the street data.	numeric	3	District_Ref (E3.4.5)	No
ROAD_STATUS_CODE	Road Status as defined within the DETR Data Capture Codes.	numeric	2	DETR Data Capture Codes	No
SWA_ORG_TYPE	The code indicates the nature of the interest of that the organisation has in the street. Defined within the DETR Capture codes.	numeric	1	DETR Data Capture Codes	Yes

Field	Description	Type	Maximum Length	Value Range	Mandatory?
START_X	The X (eastings) co-ordinate of the start point. Co-ordinates are defined in metres.	numeric	7		Yes unless WHOLE_ROAD indicator =1
START_Y	The Y (northings) co-ordinate of the start point. Co-ordinates are defined in metres.	numeric	7		Yes unless WHOLE_ROAD indicator = 1
END_X	The X (eastings) co-ordinate of the end point. Co-ordinates are defined in metres.	numeric	7		Yes unless WHOLE_ROAD indicator = 1
END_Y	The Y (northings) co-ordinate of the end point. Co-ordinates are defined in metres.	numeric	7		Yes unless WHOLE_ROAD indicator = 1

There should be at least one ADDITIONAL_STREET record for every STREET (11) record.

Record Example

```
21,36500124,1,1100,1,,1,,,1,1,,,,
```

E6.4.2 REINSTATEMENT_DESIGNATION record (22)

The Reinstatement Designation record is used to supply reinstatement categories for a street. There may be one or more such categories for a street.

If REINSTATEMENT_DESIGNATION data is being transferred, there should be at least one REINSTATEMENT_DESIGNATION record for every STREET (11) record.

Field	Description	Type	Maximum Length	Value Range	Mandatory?
RECORD_IDENTIFIER	Identifies this record as a REINSTATEMENT_DESIGNATION record.	numeric	2	22	Yes
USRN	Unique street reference number.	numeric	8		Yes

Field	Description	Type	Maximum Length	Value Range	Mandatory?
STREET_REINSTATEMENT_TYPE_SEQ_NUM	Sequential number for each street for each type of reinstatement designation.	numeric	3		Yes
STREET_REINSTATEMENT_TYPE_CODE	Reinstatement Type as defined within the DETR Data Capture Codes.	numeric	2	DETR Data Capture Codes	Yes
WHOLE_ROAD	Indicator as to whether the reinstatement category applies to the whole road. 0 indicates that it does not apply to the whole road, 1 indicates that it does.	numeric	1	0,1	Yes
REINSTATEMENT_LOCATION_TEXT	Description of the location of the part(s) of the street for which this reinstatement type is applicable	string	250		Yes, unless WHOLE_ROAD indicator = 1
REINSTATEMENT_START_X	The X (eastings) co-ordinate of the start point of the reinstatement designation. Co-ordinates are defined in metres.	numeric	7		Yes, unless WHOLE_ROAD indicator = 1
REINSTATEMENT_START_Y	The Y (northings) co-ordinate of the start point of the reinstatement designation. Co-ordinates are defined in metres.	numeric	7		Yes, unless WHOLE_ROAD indicator = 1
REINSTATEMENT_END_X	The X (eastings) co-ordinate of the end point of the reinstatement designation. Co-ordinates are defined in metres.	numeric	7		Yes, unless WHOLE_ROAD indicator = 1
REINSTATEMENT_END_Y	The Y (northings) co-ordinate of the end point of the reinstatement designation. Co-ordinates are defined in metres.	numeric	7		Yes, unless WHOLE_ROAD indicator = 1

Record Example

```
22,36500125,1,4,0,"FOR 100 METRES, FROM
JUNCTION",123456,234567,345678,
456789
```

E6.4.3 SPECIAL_DESIGNATION record (23)

The Special Designation record is used to supply special designations that apply to a street. There may be none, one or more than one such designation for a street.

Field	Description	Type	Maximum Length	Value Range	Mandatory?
RECORD_IDENTIFIER	Identifies this record as a SPECIAL_ DESIGNATION record.	numeric	2	23	Yes
USRN	Unique street reference number.	numeric	8		Yes
STREET_SPECIAL_ DESIG_NUM	Sequential number for each street for each type of special designation.	numeric	3		Yes
STREET_SPECIAL_ DESIG_CODE	The type of special restriction that the record applies to (e.g. traffic sensitive).	numeric	2	DETR Data Capture Codes	Yes
WHOLE_ROAD	Indicator as to whether the special designation applies to the whole road. 0 indicates that it does not apply to the whole road, 1 indicates that it does.	numeric	1	0,1	Yes
SPECIAL_DESIG_ PERIODICITY_CODE	Numeric code describing the periodicity of the restriction.	numeric	2	DETR Data Capture Codes	Yes
SPECIAL_DESIG_ LOCATION_TEXT	Description of the location of the special designation within the street.	string	120		Yes, unless WHOLE_ ROAD = 1
SPECIAL_DESIG_ START_X	The X (eastings) co-ordinate of the start point of the special designation. Co-ordinates are defined in metres.	numeric	7		Yes, unless WHOLE_ ROAD indicator = 1

Field	Description	Type	Maximum Length	Value Range	Mandatory?
SPECIAL_DESIG_START_Y	The Y (northings) co-ordinate of the start point of the special designation. Co-ordinates are defined in metres.	numeric	7		Yes, unless WHOLE_ROAD indicator = 1
SPECIAL_DESIG_END_X	The X (eastings) co-ordinate of the end point of the special designation. Co-ordinates are defined in metres.	numeric	7		Yes, unless WHOLE_ROAD indicator = 1
SPECIAL_DESIG_END_Y	The Y (northings) co-ordinate of the end point of the special designation. Co-ordinates are defined in metres.	numeric	7		Yes, unless WHOLE_ROAD indicator = 1
SPECIAL_DESIG_START_DATE	Date on which the special designation comes into force (if it is seasonal).	date	10		No
SPECIAL_DESIG_END_DATE	Date on which the special designation ceases to be in force (if it is seasonal).	date	10		No
SPECIAL_DESIG_START_TIME	Time at which the special designation comes into force (if it has a specified time period).	time	5		No
SPECIAL_DESIG_END_TIME	Time at which the special designation ceases to be in force (if it has a specific time period).	time	5		No
SPECIAL_DESIG_DESCRIPTION	Description providing (for certain designations) additional information.	text	120		No
SWA_ORG_REF_CONSULTANT	A code for the street authority who must be consulted with regards to the special designation.	numeric	4	DETR Data Capture Codes	No
DISTRICT_REF_CONSULTANT	The code of the operational district for the authority who must be consulted with regards to the special designation.	numeric	3	District_Ref (E3.4.5)	No

Record Example

```
23,36500125,1,2,1,2,,2,,,,,,,,06:30,09:30
```

E 6.4.4 GUIDANCE ON THE CREATION OF ADDITIONAL STREET DATA

'Type 21' (ADDITIONAL_STREET) Records

'Type 21' records should be created for every street reference type 1 and 2 within the 'type 11' record set. They should not be created against individual elementary street units (ESUs) other than where these ESUs reflect a change in characteristic of the road defined within the ASD.

Multiple 'type 21' records are only allowed against a single street (with one USRN) where;

1. other authorities have an interest in that street, or

2. the street has separate parts, each having a different Street Authority, a different Operational District within a Street Authority or a different Road Status.

Streets within additional interested authorities

Where other authorities have an interest in a street there must be one 'type 21' record for the street with the SWA_Org_Type set to 'Street Authority' and then any number of other 'type 21' records with the SWA_Org_Type field set to 'Transport', 'Bridge', 'Sewer' or 'Other Interested' Authority.

'Other Interested Authority' means a highway authority who has an interest in a street within, or immediately adjacent to their own geographic boundary, but who is not the street authority for that street.

If the other authority has an interest in only part of the street then the Whole_Road flag should be set to indicate this and a detailed description of the extent should be included within the Additional_Street_Location_Text field.

Operational District batch files for other interested authorities are available on the NSG website.

Streets with a different Street Authority, Operational District or Road Status

Under these circumstances a single street may have multiple 'type 21' records with the SWA_Org_Type set to 'Street Authority', however the Whole_Road flag must indicate that each record does not apply to the whole road and the Additional_Street_Location_Text field should include a full and detailed location description. It should be possible from the description to establish, for any part of the street, which of the 'type 21' records apply.

The following are examples of how 'type 21' records can be used in different circumstances to control the way that notifications are given.

Example 1: Copy notices for Trunk Roads.

	USRN	Additional Street Sequence Number	SWA Org Ref Authority	Whole Road	Additional Street Location Text	District Ref Authority	SWA Org Ref Maintaining Data	District Ref Maintaining Data	Road Status Code	SWA Org Type	Start X	Start Y	End X	End Y
21	12345678	1	0015	1		154	3500	001	1	1				
21	12345678	2	3500	1		003	3500	001	1	8				

In this example, the Local Highway Authority wishes to receive duplicate notices for the street. The first record indicates that the Highways Agency (0015) are the Highway Authority for the street. The second is the optional record that shows the local highway authority as an 'other interested' authority.

Example 2: Copy notices for Transport, Bridge or Sewer Authorities

	USRN	Additional Street Sequence Number	SWA Org Ref Authority	Whole Road	Additional Street Location Text	District Ref Authority	SWA Org Ref Maintaining Data	District Ref Maintaining Data	Road Status Code	SWA Org Type	Start X	Start Y	End X	End Y
21	23456789	1	1440	1		002	1440	001	1	1				
21	23456789	2	7093	1		001	1440	001	1	5				
21	23456789	3	1440	1		008	1440	001	1	1				
21	23456789	4	7187	0	Canal Br	001	1440	001	1	5	REQUIRED			

This above example shows how a second 'type 21' record is used to indicate that Railtrack (7093) wish to receive copy notices for the street. The third 'type 21' record shows how copy notices for streets with structures could be directed to a separate department within the highway authority. The last 'type 21' record shows how British Waterways would like to receive copy notices on a particular part of a street. A more detailed description than shown in the example would be required. It is probably best for all parties that copy notices are sent for the whole road as if the flag is set to '1', otherwise the promoter's notice management system will prompt user intervention which could cause problems.

Example 3: Street is part of a development that is being built and at varying stages of adoption.

	USRN	Additional Street Sequence Number	SWA Org Ref Authority	Whole Road	Additional Street Location Text	District Ref Authority	SWA Org Ref Maintaining Data	District Ref Maintaining Data	Road Status Code	SWA Org Type	Start X	Start Y	End X	End Y
21	34567890	1	2900	0	W to X	004	2900	001	1	1	REQUIRED			
21	34567890	2	2900	0	X to Y	004	2900	001	2	1	REQUIRED			
21	34567890	3	2900	0	Y to Z	004	2900	001	3	1	REQUIRED			

This example shows a street that has part adopted (from W to X), part that is prospectively maintainable (from X to Y) and part that is private (from Y to Z). In the case of the first and second parts of the street, the Statutory Undertaker would automatically generate notices for both the Highway Authority and the Street Manager as required within the Act.

Example 4:   Highway Authority boundary runs down centreline of street

In this example a street is divided down the centre by the highway authority boundary. Although, by agreement, the street may be maintained by one or other of the two authorities, the gazetteer and ASD must reflect the true situation. The NSG guidelines state that each authority must create a street record ('type 11'). It is suggested that two 'type 21' records are also created within each set of ASD.

Within this example Highway Authority "A" (1050) maintains the whole of the street by arrangement with Highway Authority "B" (3450). In both sets of ASD, Highway Authority "A" is shown as the Highway Authority and Authority "B" is shown as 'another interested authority'. By creating the data in this way, whichever street is selected from the NSG by the statutory undertaker, correct notice is given.

ASD for Highway Authority "A" (1050)

	USRN	Additional Street Sequence Number	SWA Org Ref Authority	Whole Road	Additional Street Location Text	District Ref Authority	SWA Org Ref Maintaining Data	District Ref Maintaining Data	Road Status Code	SWA Org Type	Start X	Start Y	End X	End Y
21	11123456	1	1050	1		001	1050	001	1	1				
21	11123456	2	3450	1		001	1050	001	1	8				

ASD for Highway Authority "B" (3450)

	USRN	Additional Street Sequence Number	SWA Org Ref Authority	Whole Road	Additional Street Location Text	District Ref Authority	SWA Org Ref Maintaining Data	District Ref Maintaining Data	Road Status Code	SWA Org Type	Start X	Start Y	End X	End Y
21	11123456	1	1050	1		001	3450	001	1	1				
21	11123456	2	3450	1		001	3450	001	1	8				

Example 5:   Different Operational Districts on the same street.

	USRN	Additional Street Sequence Number	SWA Org Ref Authority	Whole Road	Additional Street Location Text	District Ref Authority	SWA Org Ref Maintaining Data	District Ref Maintaining Data	Road Status Code	SWA Org Type	Start X	Start Y	End X	End Y
21	32198765	1	2470	0	A to B	003	2470	001	1	1	REQUIRED			
21	32198765	2	2470	0	B to C	004	2470	001	1	8	REQUIRED			

In this example, the Local Highway Authority as an operational district boundary across the street. Part of the street falls within district 003, and the remainder falls within district 004. Situations like this should be avoided wherever possible as it is not possible for the statutory undertaker to automatically select the correct district and therefore manual intervention is required.

'Type 22' and 'Type 23' (REINSTATEMENT and SPECIAL DESIGNATION) Records

'Type 22' and 'type 23' records should be created for every street reference type 1 and 2 within the 'type 11' record set. They should not be created against individual elementary street units (ESUs).

There should be at least one 'type 22' record for each street. Where there are no designations in place against a street there is no requirement to have a 'type 23' record, however it will not be unusual for a single street to have numerous 'type 23' records where a range of different designations are in place.

The following examples of the way designation records can be constructed are based on the 'type 22' records but the principles would also apply to 'type 23' records.

Example 1: Street with single carriageway designation only.

	USRN	Additional Street Reinstatement Type	Street Reinstatement Type Code	Whole Road	Reinstatement Location Text	Start X	Start Y	End X	End Y
22	98765432	1	4	1					

Example 2: Street with multiple carriageway designations.

	USRN	Additional Street Reinstatement Type	Street Reinstatement Type Code	Whole Road	Reinstatement Location Text	Start X	Start Y	End X	End Y
22	98765432	1	3	0	X to Y	9999999	9999999	9999999	9999999
22	98765432	2	2	0	X to Y	9999999	9999999	9999999	9999999

Example 3: Street with carriageway and footway designations applying to whole street.

	USRN	Additional Street Reinstatement Type	Street Reinstatement Type Code	Whole Road	Reinstatement Location Text	Start X	Start Y	End X	End Y
22	98765432	1	4	1					
22	98765432	2	7	1					

Example 4: Street with carriageway and footway designations applying to parts of a street.

	USRN	Additional Street Reinstatement Type	Street Reinstatement Type Code	Whole Road	Reinstatement Location Text	Start X	Start Y	End X	End Y
22	98765432	1	2	1					
22	98765432	2	7	0	A to B	9999999	9999999	9999999	9999999
22	98765432	3	6	0	D to F	9999999	9999999	9999999	9999999

APPENDIX F

Guidelines for Works at or near Railtrack level crossings

F1 PURPOSE

F1.1 Safety precautions for street works and other road works carried out in the vicinity of Railtrack level crossings are described in several separate Acts of Parliament and Regulations. These guidelines have been developed in the light of experience gained following incidents where collisions occurred from traffic tailing back across level crossings, even though the work sites were a considerable distance away. The available advice has been brought together for the first time in this appendix, to provide comprehensive guidance for all those carrying out street works and other road works at or near to level crossings.

F2 SCOPE

F2.1 This appendix specifies requirements for the execution of all works at or near Railtrack level crossings. These should be identified in the National Street Gazetteer. It applies equally to undertakers, highway authorities and others who execute works at or near level crossings. Access to, or work within, other Railtrack property is subject to separate safety requirements, details of which can be obtained from the normal Railtrack contact.

F2.2 It applies to works that take place within the boundary of the level crossing, in the highway immediately in the vicinity, or some distance away where traffic may tail back across the level crossing as a result of the traffic management system employed during the works.

F2.3 It does **NOT** apply to:

a) railways not owned by Railtrack. However, the advice is equally applicable to other railway authorities. It is strongly recommended that these principles be incorporated into appropriate arrangements for works at level crossings on railways not owned by Railtrack.

b) non-public road level crossings e.g. farm access.

All relevant legislation should be taken into account when processing these works (see paragraph F3.1).

For example, undertakers may have particular powers under their enabling legislation and wayleave or easement agreements may apply in a particular case.

Undertakers must ascertain what requirements apply before discussing their proposed works with Railtrack.

c) street running tramways.

F2.4   A working party representing the Highway Authorities and Utilities Committee (HAUC) and Railtrack PLC prepared these guidelines. Railtrack is responsible for procuring the maintenance of all infrastructure assets and for the day-to-day management of operations on the railway. The provision of train services and associated activities are the responsibility of individual train operating companies.

F2.5   Regional HAUC Committees should be the first point of contact for any queries relating to policy matters or interpretation of this appendix. It is intended to post the addresses of Secretaries of Regional HAUCs on the DETR website (www.detr.gov.uk).

F3   LEGISLATION

F3.1   For works at a level crossing, those undertaking the works must comply with the reasonable requirements of Railtrack made under section 93 of the Act. For works near a level crossing, all parties concerned must comply with the reasonable requirements of Railtrack made under the Health and Safety at Work, etc Act 1974 (HASWA) and its associated Regulations (in particular the Construction (Design and Management) Regulations 1994 (CDM)).

F4   SPECIAL FEATURES OF LEVEL CROSSING WORK

F4.1   Work at or near level crossings

Works at or near level crossings may impact upon one or more of the following: -

- safety of road users, railway passengers and personnel,

- train operation,

- structural integrity of the permanent way and other railway structures,

- railway overhead traction cables, electrified third-rails and feeder or continuity cables,

- railway underground apparatus serving the railway and running parallel to it.

The undertaker, the highway authority, Railtrack and others carrying out street and other road works have a duty to co-ordinate their activities and to follow the special safety precautions which apply at level crossings. Risk Assessments with continuous monitoring are essential to safe operation of works at or near to level crossings.

F4.2 Safety of road users, railway passengers and employees

F4.2.1 Risk assessments must be carried out both before and during works at or near to level crossings in order to minimise the risk, Safe Systems of work must be in place and maintained during the works. Risk Assessments are further described in Section F7.

F4.2.2 Traffic stopping or moving slowly over a level crossing causes potential danger to road and rail users alike. Advice on traffic control is further described in Section F7.

F4.2.3 Particular attention must be paid to situations where works which, although they may be a considerable distance away from the crossing, may cause traffic tail backs over the crossing.

F4.3 Other safety issues

F4.3.1 When work is being carried out in the vicinity of overhead traction cables, electrified third rails and feeder or continuity cables, great care must be taken to avoid danger from electrocution. This is considered further in Section F8.

F4.3.2 Electronic pipe and cable location equipment can potentially interfere with railway signalling apparatus and must not therefore be used within railway land without express permission from Railtrack, who will advise on the circumstances and type of equipment which apply at each level crossing.

F4.4 Railtrack's responsibility

F4.4.1 Railtrack will decide, on the basis of the information received from those proposing to execute works, whether the works are likely to affect train operations and advise of the arrangements made. Railtrack's Special Requirements are further described in Section F8.

F4.4.2 For works at a level crossing, those undertaking the works must comply with the reasonable requirements of Railtrack made under section 93 of the Act. In view of the requirements of the Rail Regulator, timing directions given under section 93 may entail considerable delays to the project, and therefore it is recommended that consultation with Railtrack take place at the earliest possible opportunity. For works near a level crossing, all parties concerned must comply with the reasonable requirements of Railtrack made under the Health and Safety at Work etc, Act 1974 (HASWA) and its associated Regulations (in particular the Construction (Design and Management) Regulations 1994 (CDM)).

F4.4.3 The whole of Railtrack's rail network is a continuous site for the purposes of CDM, with the local maintenance contractor assuming the role of Principal Contractor. When work is being planned to take place on or in the immediate vicinity of a level crossing, and in order for permission to be given for any works to commence, Railtrack will require details of the works and the competence of those employed to carry out the works, and will advise contact details for the Principal Contractor.

F4.4.4 Railtrack must assess the possible effects of works on the permanent way (the railway track, sleeper, ballast or other foundation material) or adjacent Railtrack land, the level crossing surface, overhead catenary supports, signalling equipment and underground railway apparatus etc. and advise on the adoption of any additional measures required.

F5 DESIGNATION

F5.1 In order to assist works promoters in fulfilling the obligations set out herein, it is recommended that the location of level crossings, and where applicable an associated Precautionary Area, where special controls will apply, should be identified and publicised using the National Street Gazetteer.

F5.2 At present this information can only be added to the National Street Gazetteer by highway authorities, although future development of ETON and management of the Gazetteer may make it possible for all interested parties to update the data. Railtrack should instigate a joint assessment to be carried out by them and the appropriate highway authority of each of the relevant sites. The highway authority must then ensure that the information is entered onto the National Street Gazetteer (see paragraphs F5.6 & F5.7).

F5.3 Highway authorities should cooperate with Railtrack in following the designation procedure set out below, in order to identify each level crossing and to establish an initial footprint of streets that will comprise the Precautionary Area.

F5.4 Undertakers and other works promoters should recognise that both NRSWA and HASWA require them to consider the implications of their works and identify the effects on traffic in the vicinity of level crossings. They should therefore co-operate in the initial establishment of the Precautionary Area and its development over time.

Identification of the Precautionary Area

F5.5 Railtrack must identify individual level crossings together with their proposals for the Precautionary Area and pass to the appropriate local highway authority, indicating the position of the crossing, its type and whether the barriers are manually or automatically controlled. The street authorities must then input this information into the National Street Gazetteer ASD data.

F5.6   Railtrack, using the street authorities' local knowledge and in co-operation with them and local undertakers, will examine each crossing to identify those streets associated with it which are likely to cause traffic tailbacks to the level crossing arising from works carried out in the highway. The exercise should identify: -

(1)   each street falling wholly or partly within 50 m of each crossing when following a route from the crossing; and

(2)   each street falling wholly or partly within 200 m of each crossing when following a route leading from the crossing but not falling beyond the second junction encountered on this route but not including junctions which consist simply of a change of street name. Junctions counted in this way should require a turning movement to or from another route.

F5.7   The whole or part of each street identified above will become part of the Precautionary Area, subject to special controls, as described below. Minor modifications may be made at this stage, for example, the exclusion of one-way streets with traffic flowing towards a level crossing.

F5.8   Streets identified above, being within 50m of a level crossing will also be subject to controls on portable traffic light signals, as described in paragraph F7.3.2.

Monitoring the Precautionary Area

F5.9   The initial footprint of the Precautionary Area should be kept under review.

F6   CONSULTATION

F6.1   Street authorities have a duty under the Act to co-ordinate all kinds of works in the street. Where this duty extends to works that are likely to affect a level crossing, Railtrack must be included in the co-ordination exercise.

F6.2   Co-ordination Meetings

F6.2.1   Railtrack may be expected to attend Co-ordination meetings when they are promoting works. Railtrack may also be expected to attend when advised of proposed street works and other road works that may affect level crossings.

F6.3   Advance Consultation with Railtrack

F6.3.1   Any planned works which will take place at or near to a level crossing, or works which are likely to affect the crossing because of traffic tailbacks (usually referred to by Railtrack as "blocking back"), must be advised to Railtrack as early in the planning process as possible, but no later than one month in advance. The form shown in Annex A must be used for this purpose.

F6.3.2   Upon receipt of advance advice of proposed works Railtrack should respond as soon as practicable in order to meet with the promoter to agree the special requirements to be included in the Health and Safety Plan for the works. This meeting may take place at a Co-ordination Meeting or separately, depending on the nature and complexity of the proposed works.

F6.3.3   Confirmation of the agreed arrangements will be given to Railtrack in writing one month before the works are intended to start. Railtrack should then give its approval of the works, or otherwise, within 10 days of receipt of the confirmation and a copy of such approval, including details of the agreed arrangements, should be given to the street authority.

F6.3.4   In the case of street authorities' works for road purposes this will be the only notification required to be given to Railtrack. In the case of undertakers and others, statutory Notices under the Act will also be required. Where the level crossing has been designated on behalf of Railtrack under the Act as a Street with Special Engineering Difficulty, undertakers and others may be required to accompany the advance advice described in paragraph F6.3.1 above with a plan and section drawing showing details of their proposed works.

F6.3.5   Contacts within Railtrack will be posted on to the DETR website (www.detr.gov.uk) as soon as possible after the publication of this Code.

F6.4   Undertakers' Statutory Notices under the Act

F6.4.1   Section 93(2) of the Act requires undertakers to give notice to Railtrack of the starting date of proposed works which are to take place at a level crossing, notwithstanding that such notice is not required under section 55 (notice of starting date).

Minimum Notice Periods

F6.4.2   This Code of Practice strongly urges that notice periods are treated as the minimum periods and, wherever possible, longer notice should be given. This is particularly important in the case of level crossings, where Railtrack may have to make special arrangements ranging from the provision of railway safety cover to complete closure of the rail route and rearrangement of rail services while the works take place.

Emergency Works

F6.4.3   Where it is necessary to carry out emergency works at a level crossing it is vital that the Street Authority and local Railtrack office is contacted immediately and work is not commenced until the undertaker has been assured that all necessary safety precautions are in place.

F6.4.4   If it is necessary to use undertaker's personnel who have not been trained in Personal Track Safety to deal with the emergency, they must not be allowed to enter the track area until Railtrack nominated personnel have arranged appropriate protection and confirmed that it is safe to do so (see Section F8).

Urgent Works

F6.4.5   Where urgent works are necessary at a level crossing, 2 hours notice should be given in advance to the street authority and Railtrack and it is vital that work is not commenced until the undertaker has been assured that all necessary safety precautions are in place.

F6.4.6   Urgent works at a level crossing should always be carried out by undertakers' personnel who hold a Personal Track Safety certificate issued in accordance with Railtrack's requirements. However, in exceptional circumstances, the arrangements described in paragraph F6.4.4 above may be followed.

F6.5   Street works licences

F6.5.1   Those without a statutory right to carry out street works must be authorised by the street authority (ie the highway authority or street managers) by means of a licence before works may commence. In addition, licensees may have to comply with the requirements of other relevant authorities or owners of apparatus affected by the works. In some cases it may be necessary to settle a plan and section. It is recommended that specific reference to this guidance should be made within licences for works in the vicinity of railway level crossings. Railtrack, as the street manager at the level crossing, will similarly ensure that the requirements of this guidance are followed when licences are issued.

F7   RISK ASSESSMENTS AND TRAFFIC CONTROL

F7.1   Codes of practice and other advice

F7.1.1   The Code of Practice *Safety at Street Works and Road Works* issued under section 65 of the Act is based on the Traffic Signs Manual (Chapter 8) and is a statutory requirement for undertakers' street works. It specifies the basic requirements for signing, lighting and guarding and traffic control at street works.

F7.1.2   Chapter 8 of the Traffic Signs Manual also gives advice on traffic control at railway level crossings, which is repeated in paragraphs F7.3.1 to F7.3.9 below, with the exception of consultation and notification procedures, which are dealt with in Section F4.

F7.2   Risk Assessments

F7.2.1   Risk Assessments are a requirement of the Management of Health and Safety at Work Regulations 1999 and it is particularly important that they are carried out at railway level crossings. Arrangements made as a result of Risk Assessments should then be continuously monitored so that appropriate measures can be taken quickly if required.

F7.2.2 Risk assessments should take into account the distance of the crossing from the proposed works and the volume of traffic using the road. Particular attention must be given to the possibility of traffic congestion tailing back over a level crossing at any time during the duration of the works, for example during an all-red traffic light period imposed to allow difficult operations to take place at the work site. There have been instances where this situation has also arisen from works that have been a considerable distance from the level crossing and this possibility should always be borne in mind during works.

F7.3 A Summary of Traffic Control Measures at or near Railway Level Crossings

F7.3.1 The following paragraphs repeat the advice for Traffic Control at Railway Level Crossings given in paragraph 2.3.12.3 of Chapter 8 of the Traffic Signs Manual, modified to use terminology consistent with this guidance.

F7.3.2 There are three very important points to be remembered about work on or near any railway level crossing:

> **1. Under no circumstances should portable traffic light signals be used at works that straddle a crossing, nor to control road traffic within 50m of level crossings equipped with twin red flashing traffic signals.**
>
> For works taking place close to the level crossing or up to 50m away from it, traffic control should be by means of 'STOP/GO' signs. Portable traffic light signals may be used for works more than 50m away from the level crossing but must be under manual control so that the operator can maintain the all-red period for sufficient time to allow the train to pass. If it is considered that road traffic may block back to the level crossing, the requirement of 2 must apply.
>
> **2. Operators must never stop road traffic on the crossing.**
>
> Where works are near to, but not on the crossing, operators should ensure that traffic does not block back and stop on the crossing: care must therefore be exercised in the traffic control arrangements. If there appears to be danger of traffic blocking back, the traffic control should be moved immediately to a point on the side of the crossing opposite the works (so that traffic can be stopped before reaching the crossing). Railtrack should then be informed from the crossing or the nearest available point.
>
> **3. Whatever method of traffic control is used, it should be ensured that the crossing's own road signals are clearly visible to approaching vehicle drivers.**

Work or any associated equipment should not obscure permanent traffic signals or advance warning and informatory signs.

F7.3.3 At automatic crossings with or without barriers, street works that may give rise to congestion will normally require the special appointment by Railtrack of a crossing attendant. All train drivers will be instructed to approach the crossing with caution and the crossing attendant will ensure that the crossing is clear before any train passes over it.

F7.3.4 When street works necessitate the provision of a crossing attendant, the traffic control arrangements described in the following paragraph should be adopted if one-way working is required. Even works that do not encroach for any great distance upon the carriageway may obscure the approach barrier or the signals controlling the crossing. Traffic control will always be required in these cases.

F7.3.5 Where the works are wholly on one side of the crossing, but within 50m of it, or wherever the build-up of waiting traffic is likely to extend from the works to the crossing, the whole of the side of the carriageway from the obstruction to the far side of the level crossing should be coned off and two manually operated 'STOP/GO' sign assemblies provided. No cones, signs or signals may be placed on the crossing. The control signs should be operated in the following manner:

- when the crossing is open to road traffic, the signs should be operated together to control traffic along the length of one-way working;

- when advised by the crossing attendant that a train is approaching, both signs should show 'STOP' to traffic coming towards the crossing. The attendant will, where practicable, operate the crossing's own traffic signals (and at automatic half barrier crossings lower the barriers);

- when the crossing attendant advises that road traffic may pass over the crossing, normal alternate one-way working should be resumed. Any vehicle held in the one-way section should be allowed to clear first.

F7.3.6 When the obstruction is on the left-hand side of the road approaching the crossing, the 'STOP/GO' sign assembly on the exit side should be sited at least 25m beyond the crossing. This is to ensure that the crossing signals are not obscured and also to allow sufficient space for vehicles to return to the left-hand side of the road beyond the crossing.

F7.3.7 When advised by Railtrack that it is not necessary for a crossing attendant to be provided, the 'STOP' signs should be shown in both directions, as described above, as soon as the amber lights in the road traffic light signals at the crossing show. Normal alternate one-way working may be resumed as soon as the red road traffic signals have stopped flashing.

F7.3.8 Street works at level crossings with manually controlled barriers or gates do not normally require a specially appointed crossing attendant. The barriers and gates and, where provided, associated road signals are controlled by a Railtrack employee, either in an adjacent cabin or remotely located and controlling the crossing with the aid of closed-circuit television. The temporarily manually operated 'STOP/GO' signs must show 'STOP' in both directions as soon as requested by the Railtrack employee controlling the crossing, or as soon as the amber lights at the crossing first show. Normal one-way working may be resumed as soon as the gates are opened or barriers lifted and, where provided, the road traffic signals extinguished.

F7.3.9 Trains are required to approach most open crossings at a slow speed. The highway approaches to open crossings are signed with 'GIVE WAY' signs and Open Level Crossing Plates. Telephones are not normally provided. If works have to be undertaken very close to such a crossing and 'STOP/GO' signs are used, the operators must keep a constant watch and stop all road traffic whenever a train approaches. They must ensure that road traffic is never stopped on the crossing. Care must be taken to ensure that the works do not obstruct the 'GIVE WAY' signs. Even though positive control of traffic at the works is required, road users will still be required to observe the permanent 'GIVE WAY' signs. Care must therefore be taken to ensure that no ambiguous instructions are given to road traffic.

F8 RAILTRACK'S SPECIAL REQUIREMENTS FOR ALL WORKS AFFECTING THE PERMANENT WAY

F8.1 Railtrack's Special Requirements must be followed in all cases where works are to take place at a level crossing and should be included in any contract let in connection with the works. The current version of this document may be obtained from Railtrack, who will also be able to give advice during the planning of works.

F8.2 Railtrack has established a robust safety regime to ensure that personnel working on or near the railway do not come into any danger from train movements. Where access will be required closer than 3 metres from the rails of a railway which is open to rail traffic (as will inevitably be the case at level crossings), it will be necessary for each individual to have been trained in personal track safety in accordance with Railtrack requirements. While on site they must be in possession of a current Certificate of Competence in Personal Track Safety. In exceptional circumstances it may be permissible for staff who do not possess a Competence Certificate to work within the confines of a level crossing (see paragraphs F6.4.4 and F6.4.6).

F8.3 Railtrack may decide where it is necessary to arrange the provision of certificated 'Lookouts' and a 'Controller of Site Safety' to be present on site for the duration of the works.

F8.4   There is no provision in the Act for the recovery of costs for special arrangements such as those described and costs incurred by each party will therefore lie where they fall.

F8.5 Controller of Site Safety

F8.5.1   Railtrack will appoint a suitably qualified person as the Controller of Site Safety. It is his responsibility to establish a safe system of work for personnel with regard to railway risks. He will decide which risk category applies at the particular site, how many Lookouts may be necessary and whether or not personnel working at the site are required to possess Certificates of Competence in Personal Track Safety.

F8.5.2   Work must not commence until the Controller of Site Safety is present and has given permission for it to do so, unless alternative arrangements have been agreed with Railtrack in advance. All personnel employed on the works must obey his instructions with regard to railway safety.

F8.6 Certificate of Competence in Personal Track Safety

F8.6.1   When a railway is open to rail traffic, a level crossing inevitably falls into the highest risk category - a "red zone". Personnel working in a "red zone" must hold a current certificate of competence in Personal Track Safety, issued in accordance with Railtrack requirements.

F8.6.2   If a railway has been closed specifically for the works to be carried out, it may be declared a "green zone" by Railtrack, in which case Personal Track Safety Certificates may not be required.

F8.7 Temporary Speed Restrictions

F8.7.1   In some instances it may be necessary to impose a temporary speed restriction on trains. Temporary speed restrictions take time to arrange, so discussions with Railtrack should take place as early as possible, before any formal notice is given for the works. Where a temporary speed restriction may already have been arranged for other purposes, the highway authority, undertaker or other person should consider rearranging the timing of its works so as to be able to make use of it, thereby minimising disruption to rail traffic.

F8.8 Trenchless construction

F8.8.1   Where trenchless construction is to be employed beneath the railway track, the method must be approved by Railtrack. A temporary speed restriction may be imposed on trains and Railtrack may require to oversee works in progress. Pipe bursting techniques will require special consideration by Railtrack, due to possible effects on the permanent way.

F8.9 Works on railway land which do not affect the structure of the permanent way

F8.9.1 These are works which do not affect the track or its surrounding land, such as works at a manhole, erection of poles and wires, pressure testing pipes which do not pass under the track or excavations several metres away from the track.

F8.9.2 The Controller of Site Safety will categorise the site, as described in paragraph F8.5.1 above, and advise of any necessary precautionary measures.

F8.10 Works outside the confines of a level crossing, using existing ducts which pass beneath the railway.

F8.10.1 These works may proceed without special early notification to Railtrack. However, Risk Assessments and Traffic Control as described in Section F7 must be followed.

F8.11 Use of mechanical excavators and cranes near the railway

F8.11.1 Railtrack will advise of any necessary precautionary measures whenever cranes, mechanical excavators, vehicles or other construction plant is to be operated close to railway traffic. The purpose is both to protect trains from accidental contact with the plant and to protect operators from the likelihood of death or serious injury arising from contact with live traction equipment and trains on the railway.

F8.11.2 Railtrack may require the submission of plant operating diagrams and/or method statements, prepared by the plant operator and approved by the undertaker, highway authority or other person carrying out the work, to enable it to decide the magnitude of any potential problems. These diagrams and method statements should take into account the possible results of machine failure, structural failure or uncontrolled operation of the plant.

F8.12 Electrification continuity cables

F8.12.1 In areas where the railway is electrified using the third rail system particular care must be taken to ensure that continuity cables, which connect the ends of the live rail to maintain electrical continuity, are not damaged whilst excavations are being undertaken. In most cases these will be laid parallel to the rails at, or just below, the adjacent ground level.

F9 REINSTATEMENT OF THE ROAD AT LEVEL CROSSINGS

F9.1 Because of the interaction with the permanent way, special requirements may apply to the reinstatement of roads and road structures at or near level crossings. Railtrack must therefore be consulted and a specification agreed, which may be carried out under their control. It is possible that there may be an arrangement already in place with the local street authority.

F10 NEW WORKS NEAR LEVEL CROSSINGS

F10.1   When undertakers are proposing to install new supplies, routes should wherever possible avoid traversing level crossings. Where avoidance is not possible, trenchless methods of installation should be considered.

ANNEX A.

SAMPLE PROFORMA FOR ADVICE OF STREET WORKS

ADVICE OF INTENDED WORKS AT OR NEAR A RAILWAY LEVEL CROSSING

To:(Name) _____ From: (Name) _____
(Company)_____ (Company)_____
(Address) _____ (Address)_____

Tel No: _____ Tel No: _____
Fax No: _____ Fax No: _____

Date: _____ Sender's Ref No: _____

PRELIMINARY ADVICE

Works are proposed at/near the level crossing at:
_____
(street/road name, railway line)

Details of the works are given in the attached description/ plan and section. (delete as appropriate)

Intended start date of works_____

Likely duration of works _____

ACKNOWLEDGEMENT OF RECEIPT OF ADVICE (by Railtrack)

Details of proposed works, Ref No: _____ has been received.

The person dealing with the proposal is:
(Address) _____ Tel No: _____
_____ Fax No: _____

The proposal has been given the Railtrack Ref No: _____

(delete as appropriate)
The works may proceed.
The works may proceed subject to receipt of a 7 day Notice confirming the start date.
The works may proceed subject to Railtrack's Special Requirements attached.
The works are not yet approved and we will contact you again by (date)_____

Signed for Railtrack _____ Date _____

APPENDIX G

Guidelines for Planning, Installation and Maintenance of Utility Services in proximity to highway and other engineering structures

G1   INTRODUCTION

G1.1   This Appendix has been modified from a document prepared by the CSS (formerly the County Surveyors' Society) in consultation with HAUC. It is addressed to undertakers, also contractors, engineers, developers, planners and others involved in excavating the highway, particularly for the installation and maintenance of underground apparatus in the street in close proximity to highway structures, but would apply equally to any other engineering structure that might be affected by the works. It reiterates the importance of prior local liaison and consultation as a means of avoiding subsequent problems. Although this Appendix has been prepared by the CSS primarily for the protection of structures owned by highway authorities, the same principles apply to structures owned and maintained by other authorities, such as Railtrack, London Underground, the British Waterways Board and others, and therefore all references in this Appendix to "highway structures" should be deemed to apply equally to structures associated with the highway but owned by other authorities.

G1.2   Purpose

G1.2.1   The purpose of this Appendix is to recommend ways in which the risk of damage to highway structures from street works may be averted, firstly by providing information on how to identify structures, then to advise on safety measures to avoid damage occurring. It also seeks to promote uniformity of approach amongst street and bridge authorities and undertakers towards the provision of services across or alongside highway structures. It reminds undertakers of the special statutory arrangements already in place for many of these structures and draws attention to the presence of others which may nevertheless be at risk of damage and require special care to be taken during street works.

G1.2.2   In order to avoid damage to highway structures, personnel engaged in street works should be able to:-

  (a) appreciate the likelihood of highway structures being affected by street works, the need for special care to be taken and the damage which can easily result from a moment's carelessness or ignorance.

  (b) recognise types of highway structure and be aware of the problems and damage that can arise if the correct procedures are not followed and also be able to recognise parts of the structure that may be exposed by excavation during street works.

(c) be aware that highway structures designated as being of Special Engineering Difficulty (SED) are subject to special statutory arrangements which set down the procedures which street authorities and undertakers need to follow before street works begin.

(d) be aware that other highway structures which may not have been designated as in (c) above, nevertheless require special care and consultation during street works.

(e) be aware that traffic management arrangements during street works may redirect traffic onto weak areas of a structure.

G1.3 Scope

G1.3.1 This Appendix should be considered when new services are to be provided or when apparatus is to be exposed or maintained in proximity to any existing highway structure.

G1.4 Background

G1.4.1 Highway structures form an essential part of the highway infrastructure that require special consideration and care when work is undertaken in the street.

G1.4.2 Society expects the provision of services (electricity, gas, water, sewerage, telecommunications etc) each of which requires an extensive distribution network.

G1.4.3 Space is often very restricted in and adjacent to highway structures and services may have to share a congested service bay within the structure. In order to satisfactorily install, operate and maintain services in such situations, the needs of the structure owner, street authority and undertakers must be jointly understood and appropriate precautions taken to minimise the risk of damage.

G1.4.4 As with all guidelines, the interpretation and application of this Appendix should be tempered at all times by good engineering judgement. The emphasis throughout is on the need for local liaison and consultation.

G2 HIGHWAY STRUCTURES

G2.1 Highway structures may be of various different forms, sizes and degrees of importance. They all need great care to be taken by those executing works in their vicinity. They may be composed of a variety of materials, e.g. brick, stone, concrete, steel, wrought or cast iron, timber or a combination of these and other materials.

Appendix G — Street Works near Highway Structures

They include, inter alia: -

> bridges (road, rail), flyovers, underpasses, subways, viaducts, aqueducts, culverts, cattle-creeps, cellars, sign-gantries, tunnels, pipes, corrugated steel pipes, footbridges, safety barriers, retaining walls, high mast lighting columns and reinforced earth structures.

G2.2 Highway structures are expensive and intricate, requiring great care and attention on the part of those installing services or repairing apparatus in their vicinity. What might appear to be minor and insignificant damage to a vital structural element may affect its safety and durability and thus substantially reduce the serviceable life of the structure. Such minor damage may remain undetected for years, but the consequences and the ultimate cost of repair could be severe.

G2.3 Modern highway structures have a design life of 120 years. Most are unique, though a large proportion of structures will fall into one of a small number of structural types. Using a few basic principles and working methods will help to safeguard their structural integrity and preserve them from damage. Annex 1 shows typical structure types and restricted zones that may be designated by the owning authority.

G2.4 Many are either scheduled ancient monuments or listed buildings, having protection under Acts of Parliament against unauthorised works that may damage their archaeological importance or special architectural or historic interest. See Section G4.

G2.5 Particular care must be taken with the reinstatement of 'high amenity surfaces' on or adjacent to highway structures designated as scheduled ancient monuments or listed buildings or located in a conservation area where their contribution to the special architecture or historic interest may be very significant. The HAUC Specification for the Reinstatement of Openings in Highways (S8.3(i)) gives the reinstatement requirements for high amenity surfaces.

G2.6 Some structures are protected by a waterproof membrane to combat the effects of corrosive de-icing salts which may penetrate the road surface. The membrane may be mastic asphalt, rubber or polymerised sheet or a thin spray-on layer. Even minor damage to this during installation of a service can cause the problems highlighted in G2.2 resulting in serious consequences to the integrity of the structure.

G2.7 Similarly many structures have movement joints either at or below the carriageway surface to accommodate expansion and contraction. There are many different types of joint: some buried, some exposed, some open and some sealed to prevent ingress of water. Minor damage to these can also have serious long-term consequences to the integrity of the structure.

G2.8　During the installation of services and maintenance of apparatus, the type of plant and equipment used for excavation and breaking out may potentially be very damaging to structures and their components unless operated with extreme caution.

G2.9　During construction of a new bridge or major maintenance work to an existing bridge, effective planning and liaison between street authorities and undertakers will often prevent future disruption and possible damage if additional ducts are incorporated within the structure at an appropriate stage. Such arrangements are likely to be in the interest of all parties and are in accordance with the principles embodied in the Code of Practice *Measures Necessary where Apparatus is Affected by Major Works (Diversionary Works)*.

G2.10　When planning a traffic management scheme for street works at or adjacent to a highway structure it is important to discuss signing, lighting and guarding arrangements with the street authority. Apart from following the requirements of the Code of Practice *Safety at Street Works and Road Works*, it is important to ensure that diverted traffic is not directed on to weak parts of the structure. It may also be necessary to erect screens to avoid debris falling through or over bridge parapets. It is also important to ensure that cables linking sets of temporary traffic signals used in connection with the works are not allowed to sag over parapets and touch live overhead rail traction cables.

G3　PRIOR CONSULTATION

G3.1　Section 88 of the Act imposes an obligation on an undertaker proposing works affecting the structure of a bridge to consult the bridge authority concerned (which may not necessarily be the street authority) before giving the usual section 55 notice and to comply with the reasonable requirements for safeguarding the structure. Annex 2 provides an example of a suitable consultation form and shows the information the bridge authority may require.

G4　ANCIENT MONUMENTS AND LISTED STRUCTURES

G4.1　Many highway structures are scheduled ancient monuments or listed buildings and protected against unauthorised works that may damage their archaeological importance or special architectural interest. They range from milestones and mileposts to extensive medieval causeways and bridges. Great care is needed to avoid damage to these structures and almost invariably specific consent is required before work on them may be undertaken.

G4.2　In some cases, from the nature of the structure, its listed status will be reasonably apparent, in others the undertaker may know of it from previous experience. Additionally, except in those cases where prior notice is not required, notification will provide the street authority the opportunity to advise the undertaker of the presence of a listed structure. These cases highlight the merit of early liaison and consultation between all parties in order to avoid delays and the possible contravention of the legislation.

G5 SPECIFICATION FOR THE REINSTATEMENT OF OPENINGS IN HIGHWAYS

G5.1　Undertakers are under a duty to carry out their works to prescribed standards. The Specification provides guidance on excavation, backfilling and reinstatement of the highway.

G6. RESPONSIBILITY FOR DAMAGE

G6.1　Under section 82 of the Act an undertaker must compensate a street authority, other undertaker or any other relevant authority for loss caused by the execution of street works. This obligation is subject to the proviso that the authority itself has not, by negligence or misconduct, contributed to the loss. These provisions should encourage all parties to liaise and co-operate fully in all cases where it is known that sensitive structures may be at risk from street works.

G7 RECOGNITION OF STRUCTURES

G7.1　Vigilance is required in the planning and execution of work. In the case of streets formally designated under section 63 as having special engineering difficulties the Regulations require details to be recorded by the street authority in the street works register. Where there is no designation, the undertaker may be aware of a structure from previous experience or local knowledge. In such instances he will be aware of the need for care in carrying out the works and can advise operatives and contractors accordingly.

G7.2　Many highway structures are large and easily recognisable but a great number are not apparent to the casual observer. Cellars, culverts and tunnels are frequently not visible from the road or verge and it is not unknown for excavation to damage underpasses or bridges without operatives becoming aware.

G7.3　A highway authority should be able to provide the location of highway structures in its ownership of which it is aware. It may also be able to provide sources of other information held on privately owned structures. Reference may also be made to other available information such as Ordnance Survey plans and records of previous installations.

G7.4　It is important that notification procedures are correctly followed. Unclear or inaccurate notices may mean that the street authority is unable to spot a potential risk to a structure or, at best, unable to determine how a structure may be affected by the proposals.

G7.5.　Features to look for include coalholes, lower ground floors, basement accesses, light wells, manholes, or gardens at a lower level than the road. In rural situations, low points may indicate probable culvert locations where watercourses pass under the highway. In hilly terrain, retaining walls may be found which both support the road and adjacent land higher than the highway.

G7.6 Undertakers are reminded of their responsibility to recognise when their proposed works will affect a structure and to consult the bridge authority before giving notice.

G8. DAMAGE TO HIGHWAY STRUCTURES

G8.1 This Appendix sets out the procedures and precautions that should be taken in order to avoid damage to highway structures during street works. It also stresses the fact that apparently minor and insignificant damage to a vital structural element may substantially reduce its serviceable life and seriously affect its safety.

G8.2 Even after taking all procedures and safety precautions into account, there may be occasions when damage nevertheless occurs. When this happens it is absolutely essential that the authority owning the structure is advised of the damage without delay so that timely repairs may be carried out. Minor damage can sometimes remain undetected for years, whilst the serviceable life and safety of the structure will deteriorate and repairs, when they are finally made, will undoubtedly prove very much more extensive and costly than if carried out immediately.

ANNEX 1

Typical Structure Types and Restrictions

1. Figures 1 to 11 show some common types of highway structures that may be encountered, the terminology used, a typical Restricted Zone that may be designated by the owning authority, typical locations of bridge waterproofing membranes and some of the constraints that the authority may place on the installation of services within it. These are not exhaustive but purely indicative.

2. Figure 1 shows a typical Restricted Zone that may be designated by a bridge authority. It will normally cover the entire width of the street and its length will extend 2 metres beyond each end of the parapet or rail. However, the extent of a designated Restricted Zone is subjective and all parties should adopt a flexible approach when considering individual circumstances to reduce the risk of damage to the structure during street works.

3. All dimensions in Figs. 1 to 11 are in millimetres.

Figure 1 Typical plans on restricted zones at structures

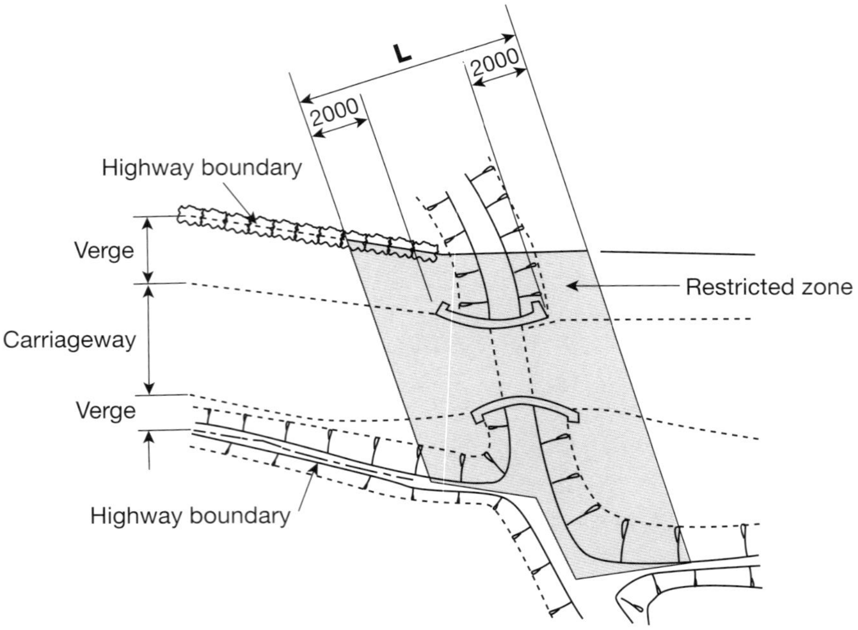
L
2000
2000
Highway boundary
Verge
Restricted zone
Carriageway
Verge
Highway boundary

SINGLE SPAN ('OLD' ARCH OR CULVERT)

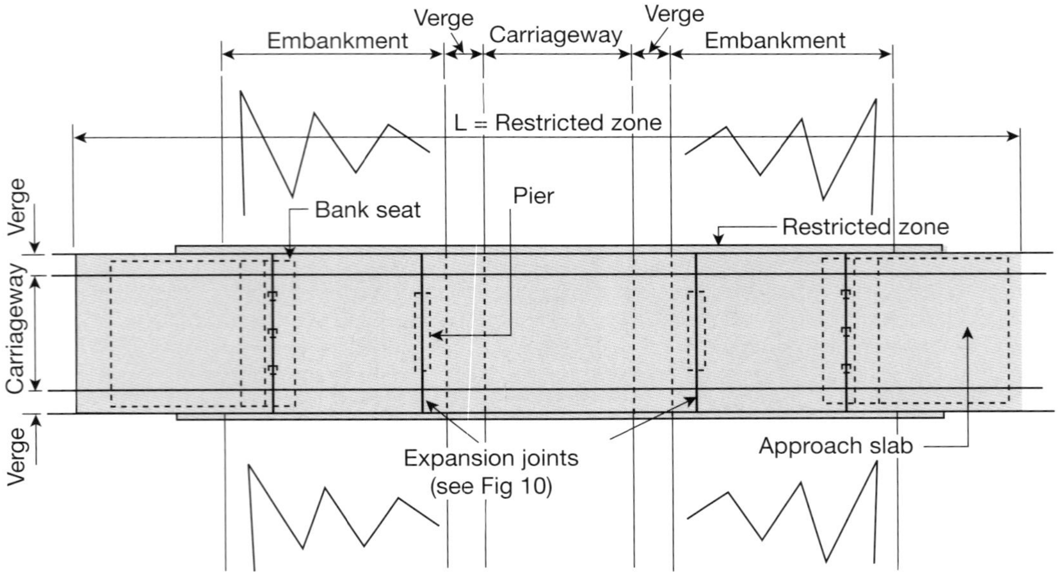
Verge
Verge
Embankment
Carriageway
Embankment
L = Restricted zone
Bank seat
Pier
Restricted zone
Verge
Carriageway
Verge
Expansion joints
(see Fig 10)
Approach slab

MULTI SPAN ('MODERN' BRIDGE)

Not to scale

Appendix G — Street Works near Highway Structures

Figure 2 Typical subway/box culvert

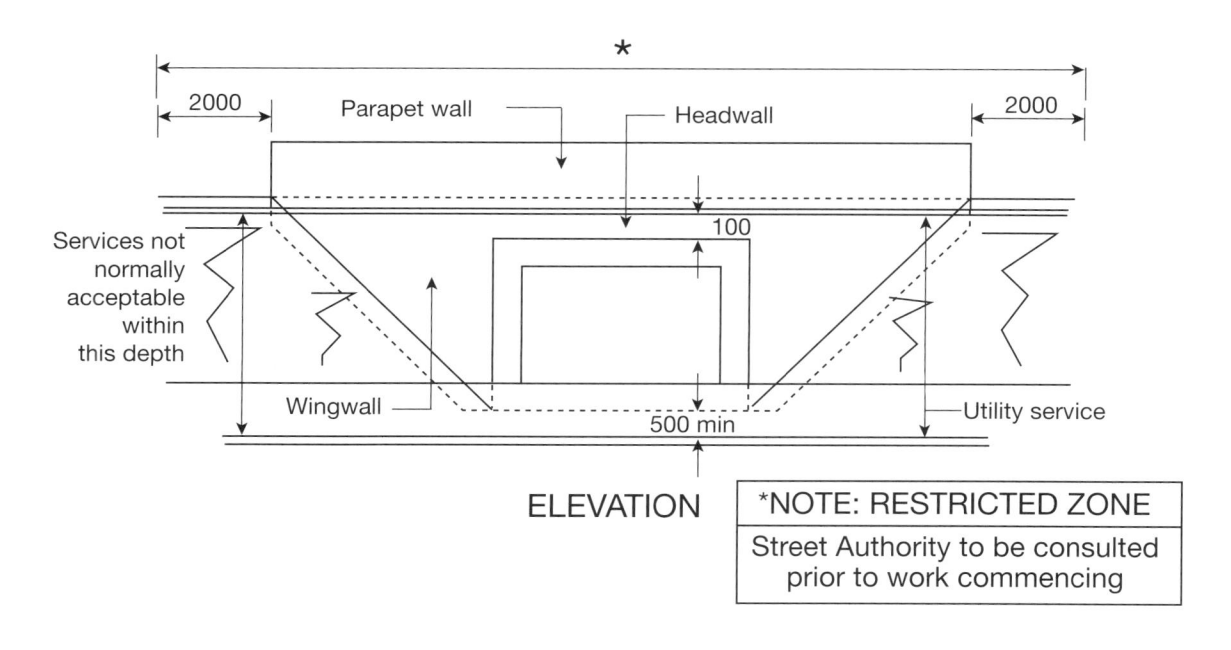

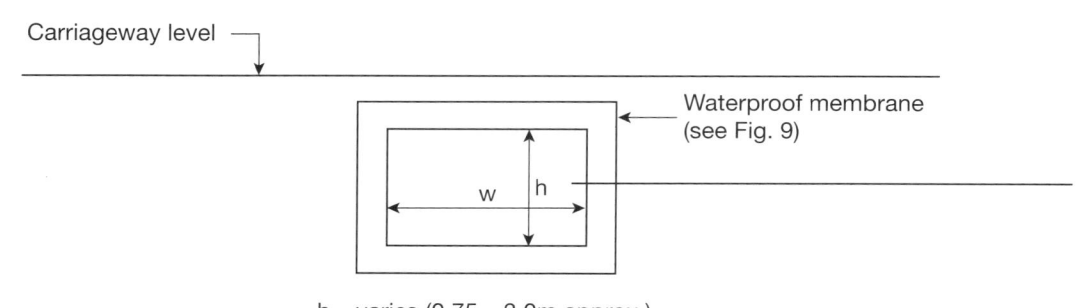

ELEVATION

- 2000 (both sides)
- Parapet wall
- Headwall
- 100
- Services not normally acceptable within this depth
- Wingwall
- 500 min
- Utility service
- * RESTRICTED ZONE

*NOTE: RESTRICTED ZONE
Street Authority to be consulted prior to work commencing

CROSS SECTION (Parallel to road)

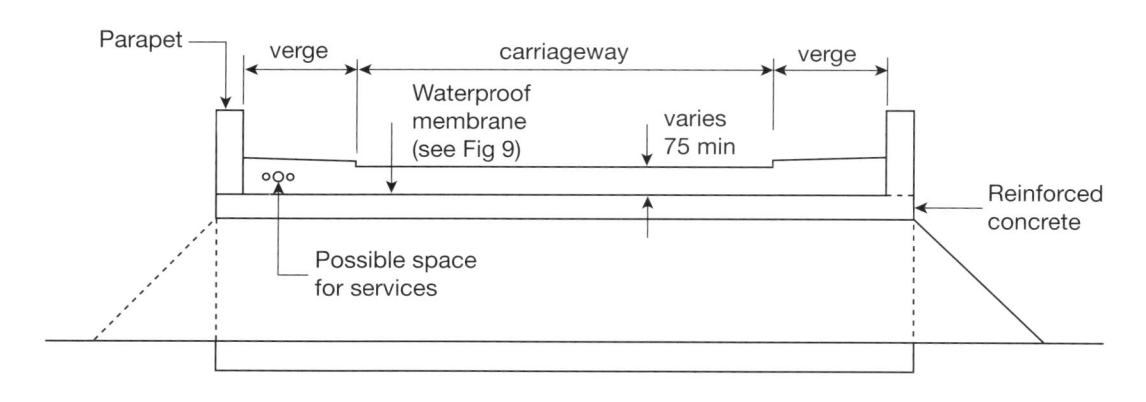

- Carriageway level
- Waterproof membrane (see Fig. 9)
- w, h

h – varies (0.75 – 3.0m approx.)
w – varies (1.0 – 4.5m approx.)

CROSS SECTION (At right angle to road)

- Parapet
- verge
- carriageway
- verge
- Waterproof membrane (see Fig 9)
- varies 75 min
- Possible space for services
- Reinforced concrete

Not to scale

Figure 3 Typical stone/brick arch or culvert

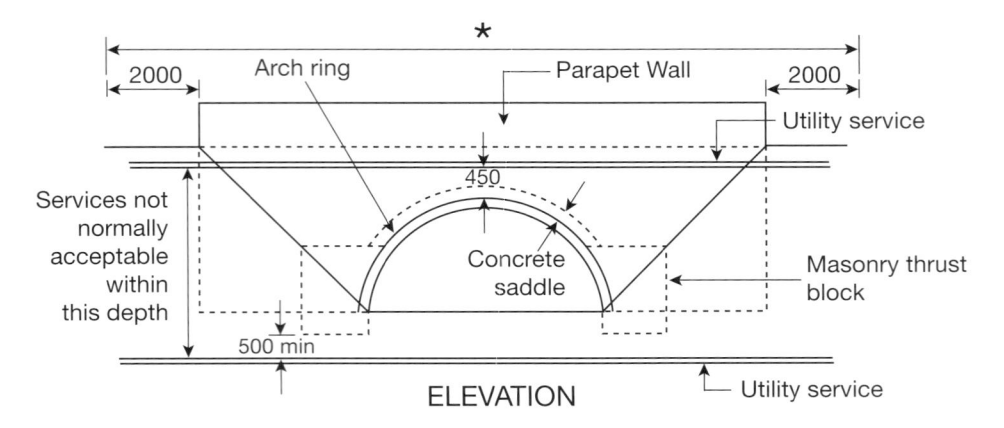

ELEVATION

- 2000 | Arch ring | Parapet Wall | 2000
- Utility service
- 450
- Services not normally acceptable within this depth
- Concrete saddle
- Masonry thrust block
- 500 min
- Utility service

*NOTE: RESTRICTED ZONE
Street Authority to be consulted prior to work commencing

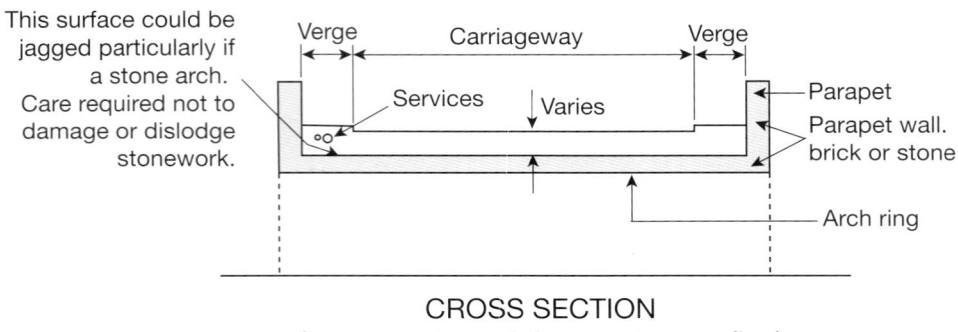

CROSS SECTION
(No concrete saddle or waterproofing)

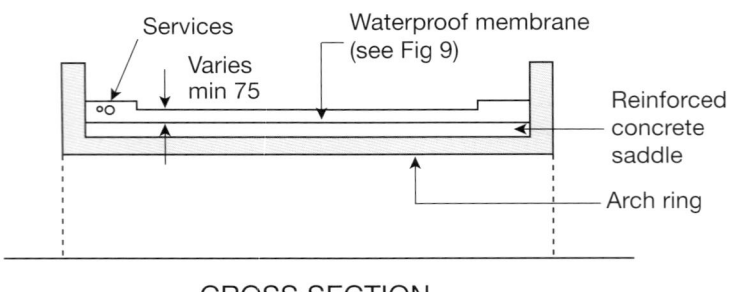

- This surface could be jagged particularly if a stone arch. Care required not to damage or dislodge stonework.
- Verge | Carriageway | Verge
- Services | Varies
- Parapet
- Parapet wall. brick or stone
- Arch ring

CROSS SECTION
(With concrete saddle or waterproofing)

- Services
- Varies min 75
- Waterproof membrane (see Fig 9)
- Reinforced concrete saddle
- Arch ring

Not to scale

Appendix G — Street Works near Highway Structures

Figure 4 **Typical single span concrete highway bridge**

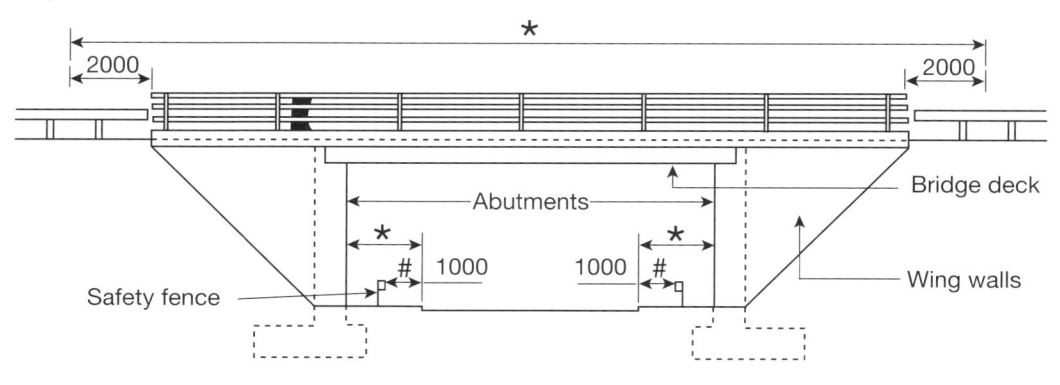

ELEVATION

#NOTE: RESTRICTED ZONE
Restricted zone adjacent to safety fence refer Fig 11

*NOTE: RESTRICTED ZONE
Street Authority to be consulted prior to work commencing

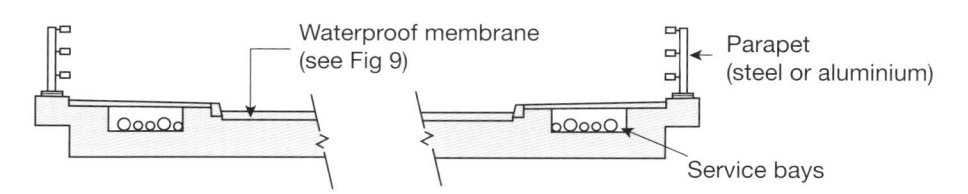

Labels on elevation: 2000, 2000, *, Abutments, 1000, 1000, Safety fence, Bridge deck, Wing walls

SECTION – WITH SERVICE BAYS

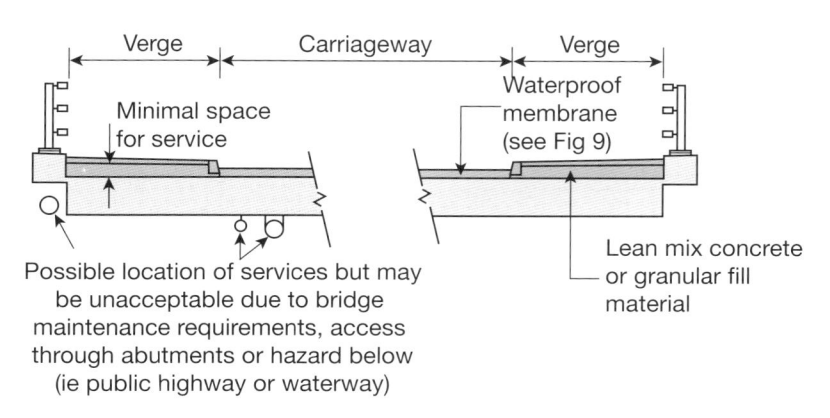

Labels: Waterproof membrane (see Fig 9), Parapet (steel or aluminium), Service bays

SECTION – NO SERVICE BAYS

Labels: Verge, Carriageway, Verge, Minimal space for service, Waterproof membrane (see Fig 9), Lean mix concrete or granular fill material, Possible location of services but may be unacceptable due to bridge maintenance requirements, access through abutments or hazard below (ie public highway or waterway)

Not to scale

Figure 5 Typical multi-span steel-concrete composite highway bridge

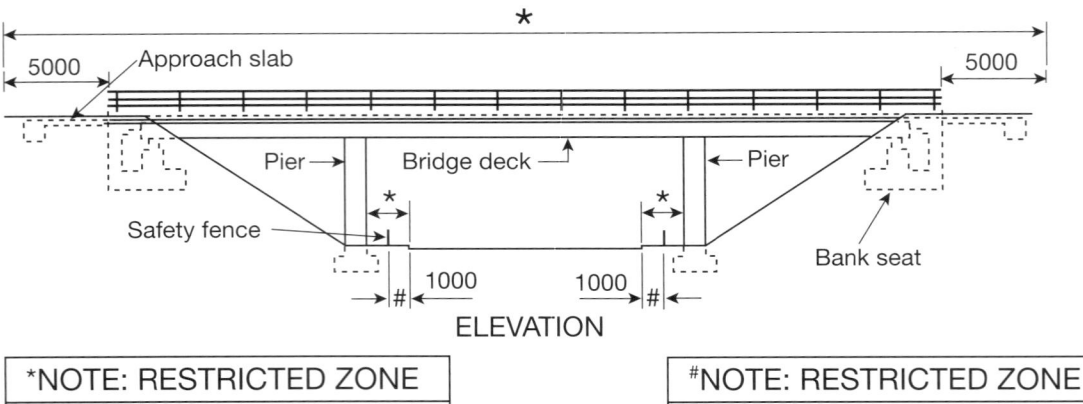

ELEVATION

- 5000
- Approach slab
- Pier
- Bridge deck
- Pier
- Safety fence
- Bank seat
- 1000
- 1000
- 5000

***NOTE: RESTRICTED ZONE**
Street Authority to be consulted prior to work commencing

#NOTE: RESTRICTED ZONE
Restricted zone adjacent to safety fence refer Fig 11

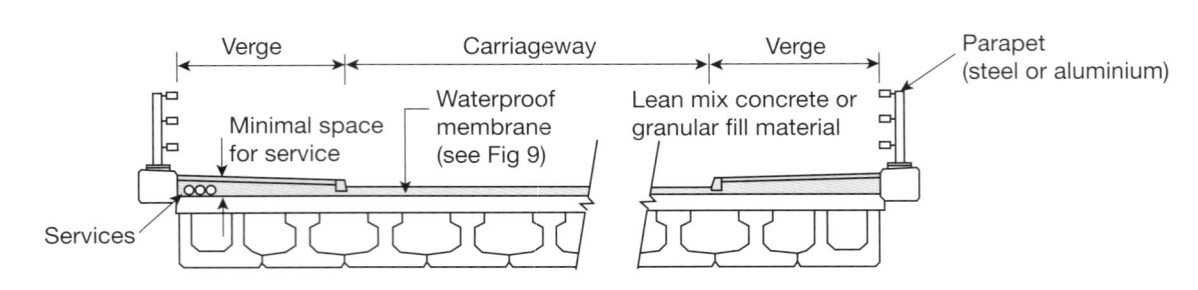

SECTION - PRECAST CONCRETE BEAMS

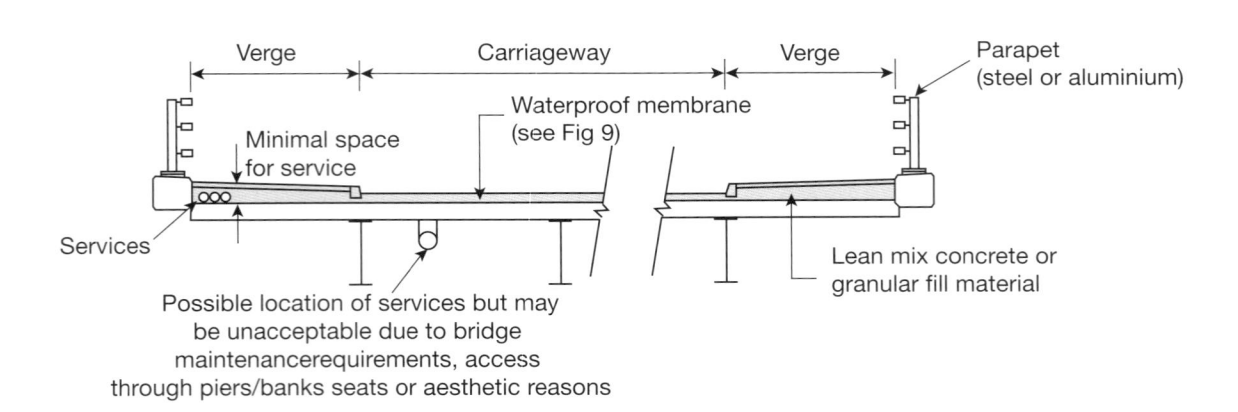

- Verge
- Carriageway
- Verge
- Parapet (steel or aluminium)
- Minimal space for service
- Waterproof membrane (see Fig 9)
- Lean mix concrete or granular fill material
- Services

SECTION - STEEL BEAMS

- Verge
- Carriageway
- Verge
- Parapet (steel or aluminium)
- Minimal space for service
- Waterproof membrane (see Fig 9)
- Lean mix concrete or granular fill material
- Services
- Possible location of services but may be unacceptable due to bridge maintenance requirements, access through piers/banks seats or aesthetic reasons

Not to scale

Figure 6 Typical retaining walls

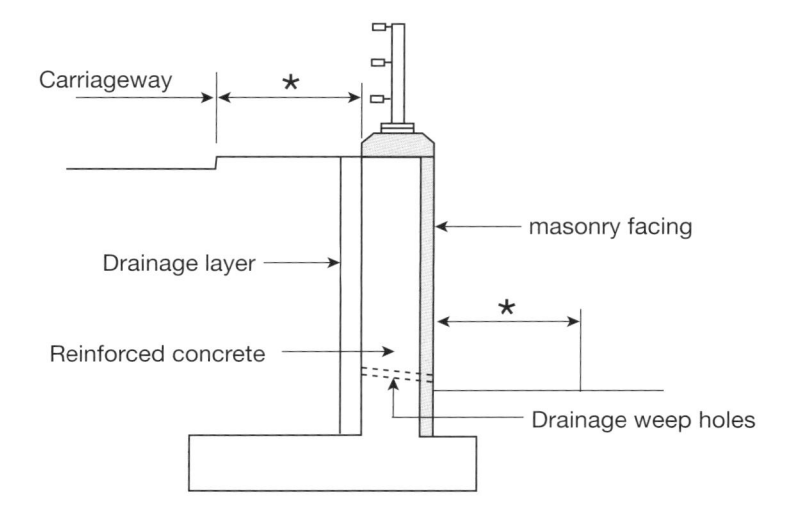

(a) REINFORCED CONCRETE WITH MASONRY FACING

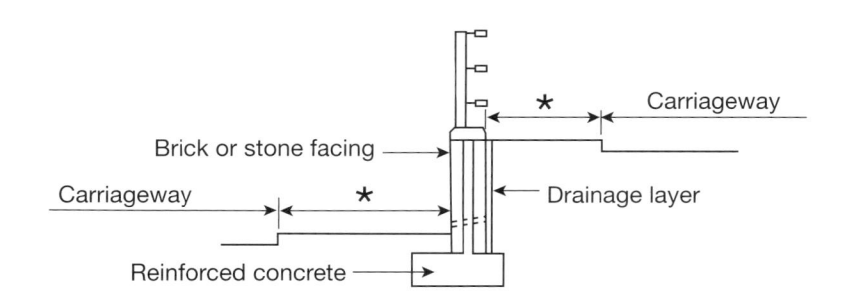

Labels: Carriageway, Drainage layer, Reinforced concrete, masonry facing, Drainage weep holes

(B) REINFORCED BRICK OR STONEWORK

Labels: Brick or stone facing, Carriageway, Reinforced concrete, Carriageway, Drainage layer

*NOTE: RESTRICTED ZONE
Street Authority to be consulted prior to work commencing

Figure 7 **Typical retaining walls (continued)**

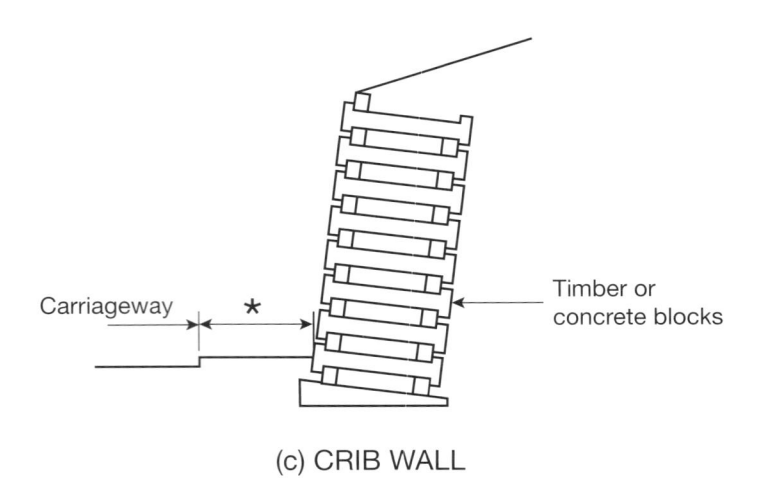

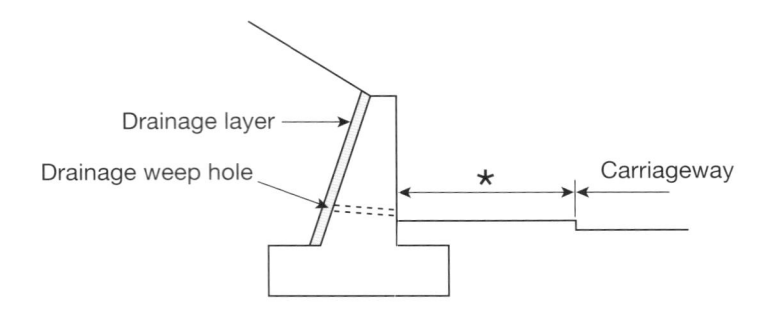

Carriageway ← * → Timber or concrete blocks

(c) CRIB WALL

Drainage layer
Drainage weep hole
* Carriageway

(d) MASS CONCRETE

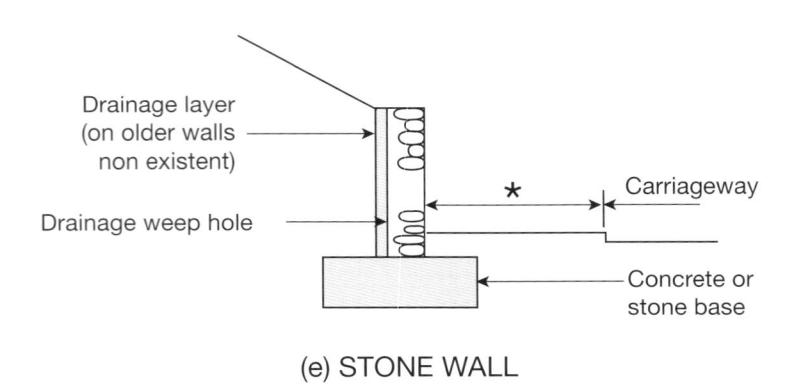

Drainage layer (on older walls non existent)
Drainage weep hole
* Carriageway
Concrete or stone base

(e) STONE WALL

***NOTE: RESTRICTED ZONE**
Street Authority to be consulted prior to work commencing

Appendix G — Street Works near Highway Structures

Figure 8 **Typical reinforced earth retaining wall**

*NOTE: RESTRICTED ZONE
Street Authority to be consulted prior to work commencing

- Verge
- Carriageway
- Steel or aluminium parapet
- Reinforced concrete anchor slab
- Precast concrete facing units
- Services may be placed in this area of verge only with great care and after prior discussion with the Street or Bridge Authority
- Metal or carbon fibre anchor straps

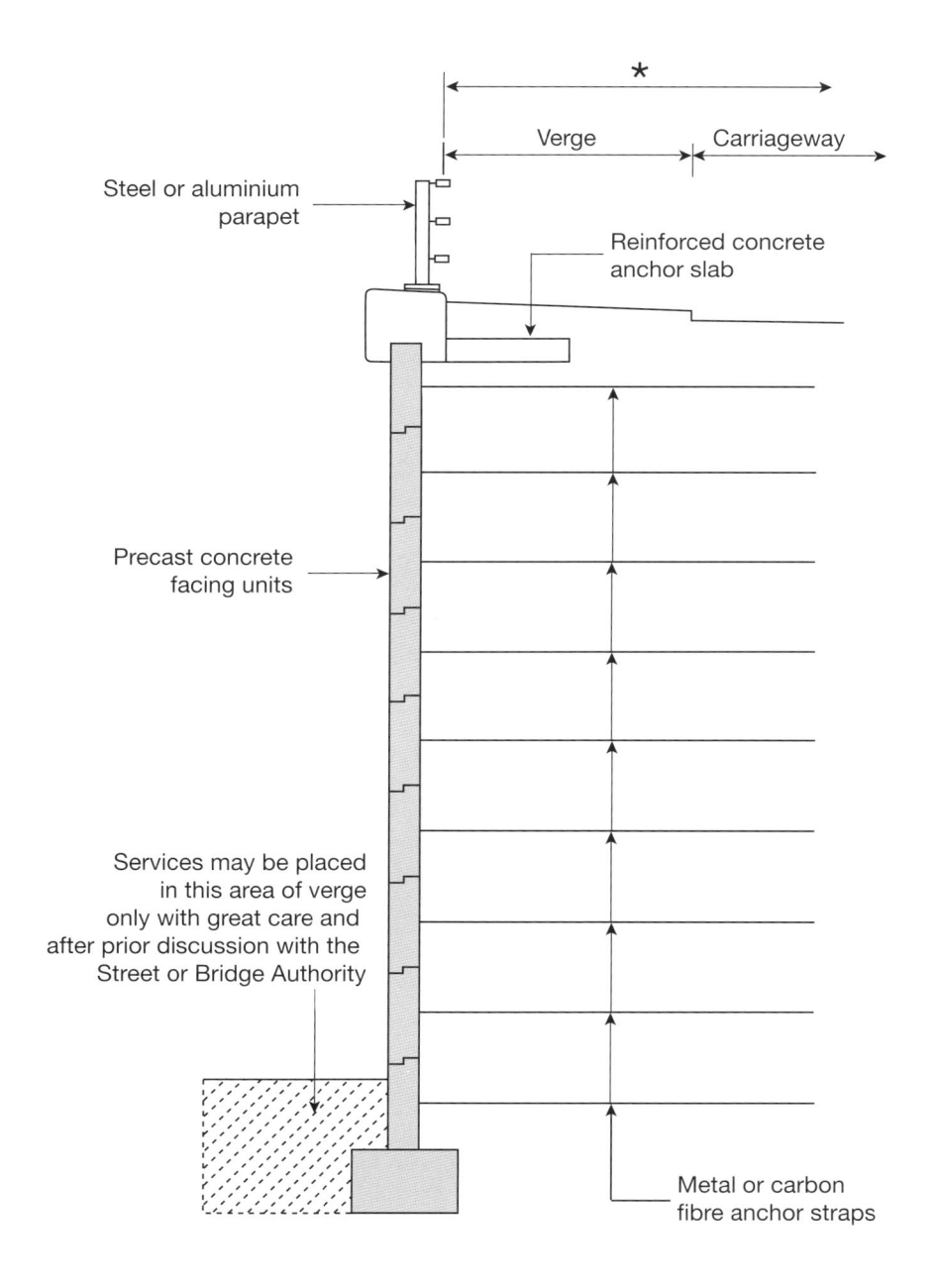

Figure 9 Typical waterproof membrane protection types

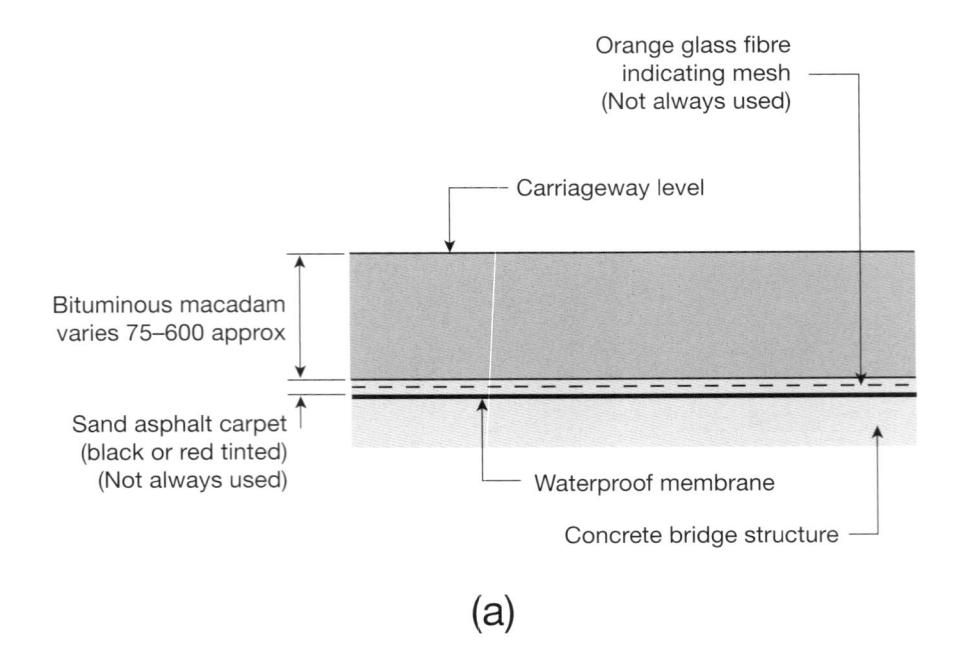

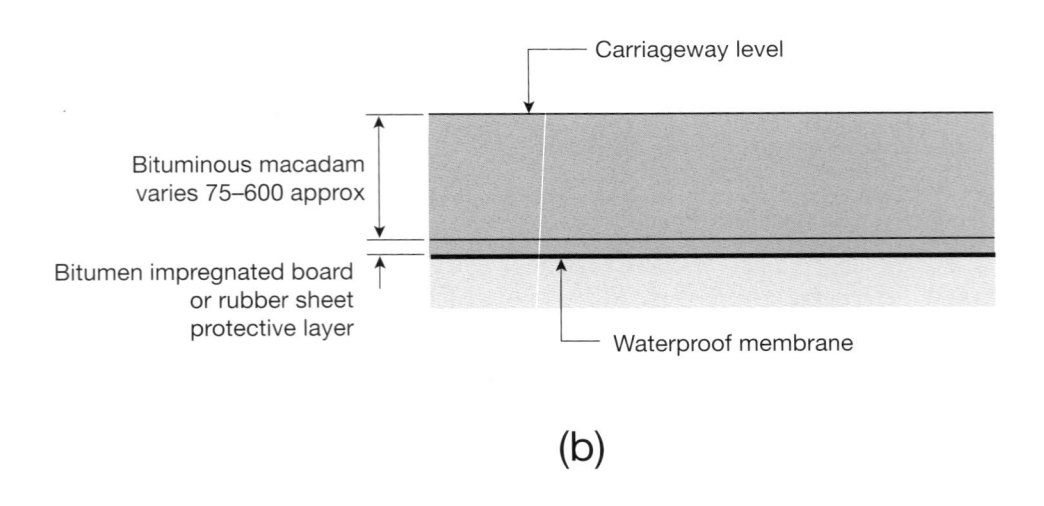

(a)

- Orange glass fibre indicating mesh (Not always used)
- Carriageway level
- Bituminous macadam varies 75–600 approx
- Sand asphalt carpet (black or red tinted) (Not always used)
- Waterproof membrane
- Concrete bridge structure

(b)

- Carriageway level
- Bituminous macadam varies 75–600 approx
- Bitumen impregnated board or rubber sheet protective layer
- Waterproof membrane

Not to scale

Figure 10 Typical expansion joints

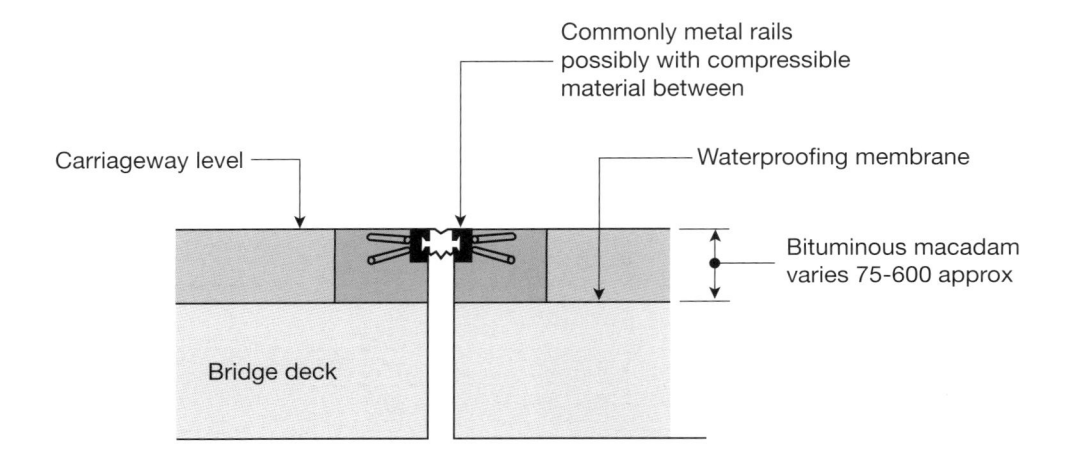

Commonly metal rails possibly with compressible material between

Carriageway level

Waterproofing membrane

Bituminous macadam varies 75-600 approx

Bridge deck

(a) EXPOSED JOINT

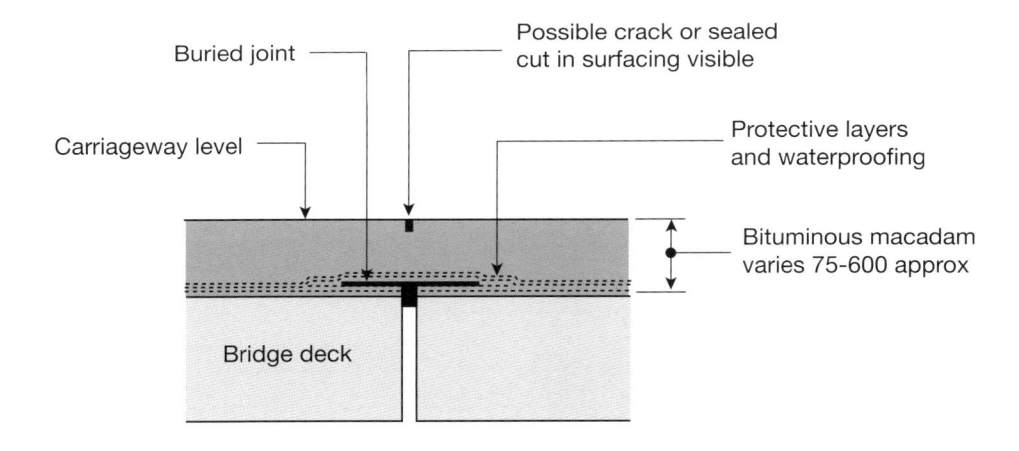

Buried joint

Possible crack or sealed cut in surfacing visible

Carriageway level

Protective layers and waterproofing

Bituminous macadam varies 75-600 approx

Bridge deck

(a) BURIED JOINT

Figure 11 **Typical safety fence**

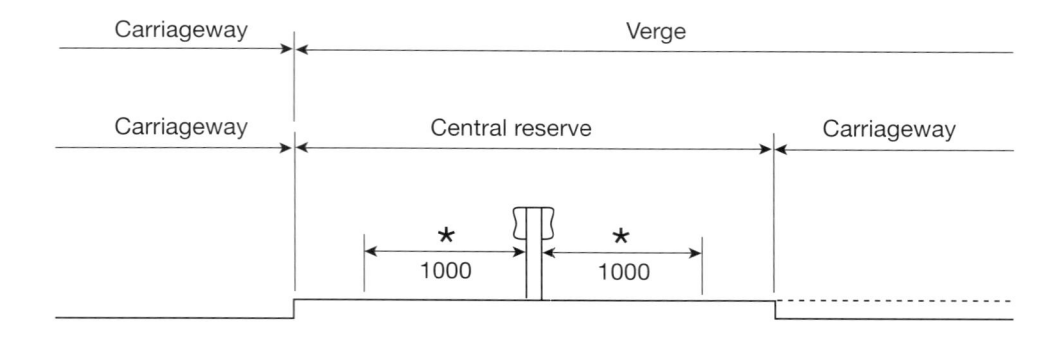

```
Carriageway  |  Verge
Carriageway  |  Central reserve  |  Carriageway
                *  1000  |  *  1000
```

*NOTE: RESTRICTED ZONE
Street Authority to be consulted prior to work commencing

ANNEX 2

Example of Consultation Form for Special Engineering Difficulty

UNDERTAKER ....

FACSIMILE TRANSMISSION
NEW ROADS AND STREET WORKS ACT 1991

To .... From ....
....
....
....
                                                  Tel No ....
Fax No .... Fax No ....
                                                  File Ref ....
Contact Name .... Contact Name ....
(if known)
DATE ....

PRELIMINARY CONSULTATION Relating to proposed works in Streets with Special Engineering Difficulty (section 63 & Schedule 4) or in the vicinity of Highway Structures (in the case of bridges, section 88)

LOCATION ....

Plan Attached   YES/NO*   (Please attach whenever possible)
Plan No   ....
Road Name/Number   ....   OS Reference ....
Bridge/Structure No   ....

DESCRIPTION OF PROPOSED WORKS ....
....
....
New/Renewal/Refurbishment*   Major/Standard/Minor*
Size and Type   ....
Proposed depth of excavation ....
Proposed depth of cover to service on completion
         ....

Expected start of works ....

BRIDGE AUTHORITY RESPONSE   File Ref ....

Please contact .... Tel No ....
Consent to proceed granted/not granted subject to* ....
....
Trial Holes required   YES/NO*   Other..
Additional details required (eg Plans, Sections, Method Statements): ....
....
....

*Delete as appropriate   Signature ....
Bridge Authority to respond within 7 working days of receipt   Date..

APPENDIX H

HAUC Terms of reference

H1 MODEL TERMS OF REFERENCE FOR REGIONAL HAUCs

H1.1 Terms of Reference

1. To report to the National HAUC through the Meeting of Regional HAUC Representatives.

2. To act as the regional focus for the National HAUC.

3. To provide a forum for discussion on issues relevant on a regional basis.

4. To act as an advisory forum in the event of unresolved local disputes and to assist in implementing conciliation and arbitration procedures.

5. To monitor the performance regionally of both the Utilities and the Highway Authorities under the New Roads and Street Works Act.

6. To refer local initiatives to HAUC for recommended adoption nationally.

7. To promote mutually beneficial good working practices in the light of National HAUC policy and current legislation.

8. To deal with specific remits from HAUC within required timescales and to report back to HAUC on these.

H1.2 Constitution

1. Representatives will be appointed by the Utilities and Highway Authorities. The client should attend meetings rather than the contractor/agent. Representatives from other organisations such as contractors may be invited to meetings with observer status at the discretion of the Regional HAUC.

2. The Chairmanship and secretariat shall rotate between the Utilities and/or Highway Authorities or by arrangement locally.

3. Both organisations will be able to submit matters for inclusion on the agenda for meetings.

4. Meetings will be held normally on a quarterly basis or at the request of the Joint Chairmen (or Secretaries if local arrangements allow).

5. Members may send substitutes to meetings and additional members may be co-opted when necessary.

6. The Regional HAUCs may set up working parties and sub-groups to work on particular subjects.

H1.3 Frequency of Meetings

Quarterly

H2 MODEL TERMS OF REFERENCE FOR LOCAL LIAISON/CO-ORDINATION MEETINGS

H2.1 Terms of Reference

To co-ordinate works to minimise inconvenience to highways users, involving:

a. consideration of both highway authorities' and undertakers' specific major projects;

b. medium term and annual programmes (both capital and maintenance) for works for road purposes and street works;

c. local policies affecting street works, including traffic management proposals;

d. reviewing performance at local level, including damage prevention;

e. street works licences

Wider issues will be referred up to Regional HAUCs.

H2.2 Membership

Representatives from any utility, or highway authority or district council carrying out street works or works for road purposes and the street authority. As the occasion requires, representatives from the adjacent street authorities, local planning authority, the police, emergency services, organisations representing disabled people, or others may attend.

H2.3 Frequency of Meetings

Quarterly